Oil and Regional Developments in the Gulf

Oil and Regional Developments in the Gulf

Edited by Rosemary Hollis

THE ROYAL INSTITUTE OF INTERNATIONAL AFFAIRS
Middle East Programme

in association with
Division of Research and Studies
Crown Prince Court of Abu Dhabi

First published in Great Britain in 1998 by
Royal Institute of International Affairs, 10 St James's Square, London
SW1Y 4LE
(Charity Registration No. 208 223)

Distributed worldwide by
The Brookings Institution, 1775 Massachusetts Avenue NW,
Washington DC 20036-2188, USA

British Library Cataloguing in Publication Data
A CIP catalogue record for this book is available from the British Library.

Hardback: ISBN 1 86203 009 X
Paperback: ISBN 1 86203 077 4

Typeset in Times by Koinonia
Printed and bound in Great Britain by the Chameleon Press Limited
Cover design by Shailesh Chavda

Contents

Contributors vi
List of seminar participants ix
Abbreviations xi
Preface xiii
Map of the Gulf and surrounding regions xvi

1. Overview *Rosemary Hollis* 1

2. Global setting and trends for oil and gas *John V. Mitchell* 12

3. Development of oil in the Gulf: the UAE in focus *David G. Heard* 34

4. Policy imperatives in Washington *Gary G. Sick* 69

5. Europe and the United States: conflicting approaches *Rosemary Hollis* 82

6. Iraq under sanctions: diminishing returns *Raad Alkadiri* 91

7. Iran and its regional relations *Edmund Herzig* 115

8. The new generation in the GCC: the case of Saudi Arabia *Mai Yamani* 136

9. Commentary and conclusions *Rosemary Hollis* 149

Contributors

Dr Raad Alkadiri is a country analyst with the Petroleum Finance Company in Washington, DC. He was formerly Middle East analyst and deputy Managing Editor with the UK-based consultancy Oxford Analytica. Prior to this, he completed a doctorate in International Relations at St Antony's College, Oxford and was teaching fellow in the Department of International Relations at the University of St Andrews from 1990 to 1991.

Dr Iskandar Bashir, although not an individual chapter contributor, has been the overall coordinator for the seminar and production of this work. He has a BA in Political Science from the American University of Beirut and a Master's degree and Doctorate in Public Administration from the Maxwell Graduate School of Citizenship and Public Affairs at Syracuse University, USA. His academic experience and professional activities cover Lebanon and the UAE. He has published a number of articles focusing on regional and international problems of development, administrative reforms and political modernization, and is the author of *Civil Service Reforms in Lebanon, The United Arab Emirates – The Status and the Future of the Union, Development in Lebanon – Second Opportunity* and *Political Modernisation.*

David G. Heard graduated in geology and physics before joining the Abu Dhabi Petroleum Company. Engaged as a petroleum engineer, he saw in the commencement of production for export of the first Murban crude oil in December 1963. He was also involved in the discovery and early development of several large Abu Dhabi onshore oilfields. He witnessed the signing of the 'Implementing Agreement' in 1978, completing the process of handing over to the government a 60 per cent shareholding and setting up the operating company ADCO to take over onshore oil operations from ADPC. Since 1980, he has been the representative in Abu Dhabi of the Abu Dhabi Petroleum Company, the 40 per cent shareholder, with the government (60 per cent), of the onshore oil concession. He received an OBE in 1990.

Dr Edmund Herzig is a Senior Research Fellow in the Russia and Eurasia Programme at the Royal Institute of International Affairs (RIIA), where he has directed a major research project on the Caucasus and Central Asia, and a lecturer in Persian Studies in the Department of Middle Eastern Studies at the University of Manchester. He is a frequent visitor to Iran and the Caucasian and Central Asian states, and is the author of a number of publications on the history and foreign relations of Iran and the Caucasus, including an RIIA research paper on *Iran and the Former Soviet South.*

Dr Rosemary Hollis is head of the Middle East Programme at the RIIA in London. Prior to taking up this appointment in 1995, Dr Hollis spent five years as head of the Middle East Programme at the Royal United Services Institute for Defence Studies. During the 1980s, she taught at George Washington University in Washington, DC, where she also gained her doctorate. The title of her doctoral thesis was 'From Force to Finance: Britain's Adaptation to Decline in Selected Arab Gulf States 1965–1985'. Her research and writing have covered the foreign policies of the United States and European states in the Middle East, Arab–Israeli relations and Gulf security issues.

John V. Mitchell is chairman of the Energy and Environmental Programme at the RIIA. He is also a research adviser to the Oxford Institute of Energy Studies, and honorary fellow of the Centre for Petroleum and Mineral Law and Policy at Dundee University. He was elected Energy Economist of the year for 1993 by the British Institute of Energy Economists. In 1994–5 he was a member of an Expert Group on Foreign and Security Policy, advising one of the European Commissioners. Recent publications include *Companies in a World of Conflict: NGOs, Sanctions and Corporate Responsibility* (ed., RIIA, 1998), *North Sea Decommissioning: Valuing the Options* (with Tracey Roberts, OLES, Oxford, 1997); *The New Geopolitics of Energy* (with Peter Beck and Michael Grubb, RIIA, London, 1996) and *An Oil Agenda for Europe* (with Stephen Peake, RIIA, London, 1994).

Professor Gary G. Sick served on the National Security Council staff under Presidents Ford, Carter and Reagan. He was the principal White

House aide for Iran during the Iranian revolution and the hostage crisis and is the author of two books on US–Iranian relations. Mr Sick is a captain (ret.) in the US Navy, with service in the Persian Gulf, North Africa and the Mediterranean. He was the deputy director for International Affairs at the Ford Foundation from 1982 to 1987, where he was responsible for programmes relating to US foreign policy. Mr Sick has a Ph.D. in Political Science from Columbia University, where he is Senior Research Scholar and Adjunct Professor of International Affairs. He is a member of the board of Human Rights Watch in New York, chairman of the advisory committee of Human Rights Watch/Middle East, and executive director of the Gulf 2000 Project.

Dr Mai Yamani is Research Fellow with the Middle East Programme at the RIIA. She studied anthropology at Bryn Mawr College, Pennsylvania, before going up to Somerville College, Oxford for her Master's degree and subsequently doctorate in social anthropology. In 1981, she lectured in Social Anthropology and Sociology at King Abdul Aziz University, Jeddah. Subsequently, she has been a research fellow at the Centre for Cross Cultural Research on Women, Oxford University, and academic adviser to the Centre for Contemporary Arab Studies at Georgetown University, Washington, DC. She lectures at the Centre of Islamic and Middle Eastern Law at SOAS and is currently conducting a major research project at the RIIA, on the new generation in the GCC states, which will provide insights on the values and aspirations of young people in the Arab Gulf states.

List of seminar participants

H.H. Sheikh Sultan Bin Khalifa Al Nahayan, Chairman, Crown Prince Court

H. H. Sheikh Saeed Bin Saif Al Nahayan (Chairman of the Seminar), Research and Studies Division, Crown Prince Court

Ahmad Umar Abdullah, Marketing Department, ADNOC

Dr. Atif Ajwa, Economic Expert, Planning Department of Abu Dhabi

Aziza Al Hamady, Manager, Department of Research and Studies, Ministry of Economics and Commerce

H.E. Abullmalik Al Hamar, Adviser, Presidential Court of UAE

H.E. Dr Yousuf Al Hasan, Adviser, Ministry of Foreign Affairs

Dr Sulaiman Al Jassim, Director for Community Relations and Manpower Development, Higher Colleges of Technology

Dr Raad Alkadiri, Country Analyst, the Petroleum Finance Company

H.E. Dr Jassim Abdulghani Al Khulufi, Deputy Ambassador, Ministry of Foreign Affairs

Dr Hasan Alkim, Chairman of Political Science Department, UAE University, Al Ain

Hashim Fawaz Al Kudssi, Treasury Senior Manager, ADIA

Mubarak Al Kutbi, Marketing Department, ADNOC

Salem Al Muheirbi, Head, Political Division, Crown Prince Court

H.E. Khalid Eid Al Muraikhy, General Director, Crown Prince Court

H.E. Ubaid Bin Saif Al Nasery, Minister of Petroleum and Mineral Resources

H.E. Naser Bin Ahmad Al Suwaidi, Director, Marketing Department, ADNOC

Dr Iskandar Bashir, Adviser, Crown Prince Court

Dr Sabri Hasanein, Economics Expert, Crown Prince Court

David G. Heard, OBE, Representative, Abu Dhabi Petroleum Co. Ltd

Dr Frauke Heard-Bey, Centre for Documentation and Research, Abu Dhabi

Dr Edmund Herzig, Senior Research Fellow, Russia and Eurasia Programme, RIIA, and Manchester University

Dr Rosemary Hollis, Head, Middle East Programme, RIIA

Dr Saeed Mushtak Hussein, Chief Technical Adviser for the UNDP Strategic Study Programme

Roy M. Jordan, Marketing Department, ADNOC

John V. Mitchell, Chairman of Energy and Environment Programme, RIIA

Sir John Moberly, Associate Fellow, Middle East Programme, RIIA, and former UK Ambassador to Jordan and Iraq

Patrick Morgan, Deputy Head of Mission, British Embassy, Abu Dhabi

Munif Othman, Senior Investment Manager, ADIA

H.E. Mustafa Soheim, Legal Adviser for the Emirate of Abu Dhabi, Crown Prince Court

Dr Mohamad Shehab, National Coordinator for the UNDP Strategic Study Programme

Professor Gary G. Sick, The Gulf 2000 Project, Columbia University

Abdullkareem Thabet, Marketing Department, ADNOC

Dr Naif Ubaid, Political Expert, Crown Prince Court

Dr Mai Yamani, Research Fellow, Middle East Programme, RIIA

Abbreviations

ADCO	Abu Dhabi Company for Onshore Oil Operations
ADGAS	Abu Dhabi Gas Liquefaction Limited
ADIA	Abu Dhabi Investment Authority
ADMA	Abu Dhabi Marine Areas
ADMA-OPCO	Abu Dhabi Marine Areas Operating Company
ADNOC	Abu Dhabi National Oil Company
ADPC	Abu Dhabi Petroleum Company
AIOC	Azerbaijan International Operating Company
bbl	(price) per barrel
BPC	Basra Petroleum Company
bpd	barrels per day
CASCO	Caspian Sea Cooperation Organization
CCGT	Combined Cycle Gas Turbine Generation Technology
CENTCOM	US Central Command
CIS	Commonwealth of Independent States
CNN	Cable News Network
CO_2	carbon dioxide
ECO	Economic Cooperation Organization
EIA	Energy Information Agency
EIU	Economist Intelligence Unit
EME	Emerging Market Economies
EU	European Union
FAO	Food Aid Organization
FSU	Former Soviet Union
GASCO	Abu Dhabi Gas Industries Company
GCC	Gulf Cooperation Council
GDP	Gross Domestic Product
HOPEC	High OPEC (production)
IAEA	International Atomic Energy Agency
IEA	International Energy Agency
ILSA	Iran and Libya Sanctions Act
IMF	International Monetary Fund

IPC	Iraq Petroleum Company
IPIC	International Petroleum Investment Company
JMC	Joint Management Committee
JODCO	Japan Oil Development Company Limited
LNG	liquefied natural gas
LOPEC	Low OPEC (production)
MEED	*Middle East Economic Digest*
MEES	*Middle East Economic Survey*
MENA	Middle East, North Africa economic conference
mta	million tonnes a year
NEDC	Near East Development Corporation
OAPEC	Organization of Arab Petroleum Exporting Countries
OECD	Organization for Economic Cooperation and Development
OIC	Islamic Conference Organization
OPEC	Organization of Petroleum Exporting Countries
OSCE	Organization of Security and Cooperation in Europe
RCC	Revolutionary Command Council
SPC	Supreme Petroleum Council
TCF	trillion cubic feet
UAE	United Arab Emirates
UNSC	United Nations Security Council
UNSCOM	United Nations Special Commission
WMD	weapons of mass destruction
ZADCO	Zakum Development Company

Preface

This work is the product of a joint undertaking by the Research and Studies Division of the Crown Prince Court in Abu Dhabi and the Middle East Programme of the Royal Institute of International Affairs (RIIA) in London. In November 1997 the Crown Prince Court hosted a symposium on Oil and Regional Developments in the Gulf, for which the RIIA provided a team of visiting speakers knowledgeable in the fields of energy economics, regional politics and international relations. The seminar afforded an opportunity for the visiting specialists to put their views on contemporary developments affecting the Gulf region as a whole, and the United Arab Emirates (UAE) in particular, to an assembled group of experts from the Emirates, including some from the ministries in Abu Dhabi and researchers from the Crown Prince Court. On the central issue of oil and the place of the UAE in the international energy sector, a professional in this field from Abu Dhabi presented his appraisal of the evolution of the industry in the Emirate and its prospects in conjunction with an overview of trends in the energy market as a whole from a visiting specialist.

The seminar gave rise to a lively discussion between the various participants, enabling a distillation of views from within and outside the region. The result is the set of essays which follows. All the contributing authors have revised their original papers, in the light of the exchanges which took place in the symposium and to reflect the passage of events since. The concluding commentary, in particular, features the concerns and questions raised by those in the region in response to the perspectives of those who view the region from an external vantage point. Consequently, this collected work represents a searching analysis of the principal factors affecting oil and related political developments in the Gulf, setting the UAE in this broader context.

As was stated by H.H. Sheikh Sultan Bin Khalifa Al Nahayan in his opening speech at the symposium, it is through dialogue between cultures that understanding is enhanced and knowledge advanced. It was with this in mind that the symposium was convened to enable a pooling of views from across cultures. As also explained by Sheikh Sultan, the endeavour was specifically designed to examine simultaneously the realms of politics and economics. The prospects for the region's energy sector cannot be understood in isolation from the interplay of political interests and the involvement of outside powers in the area.

The last decade of the twentieth century opened with a major war between Iraq on the one hand and the United States and its allies on the other. That war liberated Kuwait but left many other issues unresolved. US strategies to contain both Iraq and Iran have meant the advent of a new security structure in the area, one facet of which is an enhanced role for US and allied forces in the defence arrangements of the Gulf Cooperation Council (GCC) states. Those states nonetheless look forward to a more tranquil time of good-neighbourly relations and cooperation in the area. Meanwhile, the legacy of the 1990–91 Gulf war and its aftermath in the energy sector could not have been more pronounced. The purpose of this volume is to assess the net impact on the energy sector of the political factors operative in the region and to identify the fundamental forces for change beneath the surface.

For the UAE, as indicated by Sheikh Sultan, the twin goals of energy development and economic diversification are the prerequisites for meeting the needs of the new generation in the Emirates, as in neighbouring countries, as a new century dawns. This work attempts to set the scene for the advancement of those aspirations.

The views expressed in the essays which follow are those of the authors alone. Their enthusiasm and commitment to the whole endeavour are much appreciated. Special thanks are due to H.H. Sheikh Sultan Bin Khalifa Al Nahayan and the Crown Prince Court of Abu Dhabi for making it possible. A vote of thanks is also in order to H.H. Sheikh Saeed Bin Saif Al Nahayan, who inspired the project and offered guidance from its inception to completion, and to Dr Iskandar Bashir who coordinated the work throughout. The success of the dialogue which informed this volume derives from the contributions of all the participants in the symposium, who are listed separately, though again, they bear no responsibility for the finished product. The editor alone takes responsibility for the summary of ideas gleaned from the symposium which appears in the concluding chapter here. Finally, to the staff of the Research and Studies Division of the Crown Prince Court and of the

Royal Institute of International Affairs, in particular Raksha Thakor, administrator of the Middle East Programme, many thanks for their efforts in organizing the symposium and for helping to produce this volume.

September 1998 R.H.

Map of the Gulf and surrounding regions

1 Overview

Rosemary Hollis

Energy producers and consumers share an interest in understanding how international relations and regional developments in different parts of the world will affect both demand and supply. From the opening years of the twentieth century, the Gulf region has been a focus of interest because of its enormous oil and gas resources. As the century draws to a close, the importance of this region is even more apparent, since at least 60 per cent of the world's remaining oil reserves and a sizeable proportion of untapped natural gas supplies are known to reside there. Energy consumption worldwide continues to edge upwards, in keeping with economic development and population growth. Yet, within this broad context, the prospects for Gulf energy producers are far from certain. This is partly because of uncertainties at the global level about economic growth patterns and hence energy consumption, but also because of changing attitudes towards environmental issues, which will impact on the energy sector. In addition, the global strategic power balance has undergone a transformation with the end of the Cold War, and nowhere may the effects of this be more apparent than in the Gulf region, where the United States exercises unprecedented influence, but indigenous political developments have their own dynamics, irrespective of what governments either outside or within the region may intend or prefer.

The intention here is to examine the different forces at work and arrive at an understanding of how they interact and interweave, thereby determining the prospects for individual energy producers in the Gulf. In the chapters which follow, an assessment will be made of the global energy scene, looking specifically at trends in oil supply, demand and prices. Within this context, the case of the United Arab Emirates (UAE), and more specifically Abu Dhabi, will receive special attention as a study in how this oil-rich country has come into its own as a major producer and how its prospects continue to be shaped by unfolding economic and political developments at the international and regional levels.

Picking up on the theme of political developments at the international level, the analysis will then turn to the role of outside powers in the Gulf region, focusing on the posture and policies of the United States. In the aftermath of the 1990–91 Gulf war, the United States developed a formula for maintaining security and stability in the area predicated on defence guarantees to the Gulf Cooperation Council (GCC) states of Kuwait, Saudi Arabia, Bahrain, Qatar, the UAE and Oman, and 'dual containment' of Iraq and Iran. As the 1990s draw to a close it is clear that this policy is undergoing a rethink in Washington, in part prompted by changes inside the Gulf countries themselves, but also fuelled by pressures from America's allies in Europe. In Chapter 4, on US policy, the problems that have faced Washington in sustaining its formula of 'dual containment' are discussed and the author's expectations for change explained. Complementing the chapter on the United States is one by Rosemary Hollis on the perspectives of Europeans, which sets out how and why they have differed from Washington, specifically on the concept and implications of 'dual containment'.

Attention will then turn to the regional setting in Chapters 6 and 7, focusing on Iraq and Iran respectively. The thrust of the analysis of the situation in Iraq, under United Nations sanctions since its invasion of Kuwait in 1990, is that the cumulative effects of the embargo have been contrary to apparent US intentions and bode ill for the future tranquillity of that country, if not of its neighbours too. On Iran, the assessment is that US-imposed sanctions have contributed to a determination to avoid isolation and that the Islamic Republic's dealings with its northern neighbours, in the Caucasus and Central Asia, seem to suggest a blueprint for its more recent rapprochement with the Arab states of the Gulf. Some consideration is also given to internal political developments in Iran and what they seem to portend.

The review of regional developments culminates in Chapter 8 with a look at the new generation in Saudi Arabia. Like the other member states of the GCC, Saudi Arabia has been experiencing a population explosion, which will lead to mounting pressures on the government to restructure the economy to meet rising welfare costs and create more jobs. The analysis presented here seeks to define the factors which are shaping the attitudes and values of youth in the Saudi kingdom and thereby arrive at some understanding of the aspirations and preferences which will characterize the new generation in the coming century. The insights gleaned no doubt will have parallels in the other member states of the GCC, while the decision to focus specifically on Saudi Arabia derives from an expectation that all its neighbours as well as external powers involved in the region are likely to feel some ripple effect from developments determining the future orientation of the kingdom.

To conclude, Chapter 9 provides a summary of how all the issues discussed are viewed from within the region, specifically by observers in the UAE. It serves as a reminder that from within this setting policy priorities are inevitably different from those affecting the outside powers and even other regional players. For the United States and Europe the region is of strategic interest, and their calculations tend to reflect a pre-occupation with safeguarding their own access to energy supplies and markets. For Iraq the priority remains survival, and for Iran a major concern is to manage the transition from revolution to post-revolutionary politics, both internally and in regional relations. In Saudi Arabia, accommodation of the needs of the new generation has to be balanced with an assessment of what is possible in the current economic climate. For all the regional players, of course, economic prospects are shaped by international energy developments. Thus, the analysis offered here comes full circle and, as demonstrated in the discussion in Chapter 9, the concerns of the UAE are about determining the future of the nation in the face of all the interacting forces that affect the country's priorities and options.

The rest of this chapter describes the unfolding dynamics of Gulf oil politics and the broad themes which seem to characterize the situation on the eve of the twenty-first century.

Oil, politics and the Gulf

In the mid-1970s, some 65 per cent of crude oil production outside the Soviet Union derived from members of the Organization of Petroleum Exporting Countries (OPEC), with Gulf producers accounting for the bulk of this. A decade later, however, OPEC's share of production in the non-communist world had declined to 40 per cent. In the interim, supplies from non-OPEC countries had increased and, with the application of new technologies and improved returns on initial investments in areas like the North Sea, this trend has continued to the end of the century. Meanwhile, oil prices have slithered downwards from peak levels in the early 1980s, such that, in real terms, prices for OPEC crude in early 1998 had slipped back to their pre-oil boom levels of the early 1970s. The most immediate causes of the price slump were probably the collapse of the developing economies in Asia, and the impact on demand of the weather phenomenon *El Niño* and a mild winter in Europe. Underlying responsibility could be attributed to the continued growth in supply levels worldwide, though clearly this is not the whole story.

Demonstrably, global economic interdependence, which has attracted so much media attention in the late twentieth century, has always been operative in the energy sector. For the oil producers of the Gulf this has been apparent in a number of ways. Market forces, technological developments and climate change represent only part of the picture. Political factors have also had a profound impact on consumer choices and the fortunes of producers. Political developments in the Gulf have fuelled the perception that this is an intrinsically unstable region. The Gulf states have not been inured to the Arab–Israeli conflict, which has flared into war several times in the past half-century and remains a source of great tension. Nationalist sentiments prompted the governments of producing countries in the Middle East to take over control of the industry from the foreign oil companies in the early 1970s, heralding the major price hike that unleashed inflation in the Western economies. This and the decision of some producers to cut oil exports to countries supporting Israel in the 1973 war led Western governments and companies to adopt policies designed to reduce their exposure to oil shocks triggered by political developments in the Middle East.

Even so, there was to be another and more pronounced oil price boom at the end of the 1970s, again attributable to Middle East politics as the Islamic revolution in Iran toppled the Shah's regime and the American hostage crisis unfolded. Panic buying in the oil market and price increases imposed by hawkish members of OPEC held sway, despite efforts by Saudi Arabia to force prices down in the interests of maintaining consumer loyalty. In fact, prices did begin to slip, as economic recession in the consumer countries began to take effect, but

Iraq's invasion of Iran in September 1980 kept the crisis going as each side set about damaging the other's energy sector. In due course, however, market forces made themselves felt and by 1982 OPEC members found themselves agreeing to a quota system to keep production down sufficiently to stabilize prices. Saudi Arabia was effectively recognized as 'swing producer'.

The net effect of these developments is now well known. Nationalization of oil in the Middle East prompted the commercial companies to look elsewhere, and higher prices made it viable for them to develop fields previously considered too inaccessible to be cost-effective. By the time the price hawks in OPEC and panic reaction to the upheavals in the Gulf region had taken their toll on energy demand and the crisis had peaked, OPEC and with it the producers of the Gulf had lost their commanding market share. Whereas in 1979 OPEC had been producing 31m barrels per day (bpd), by March 1982 the organization felt obliged to agree to an overall output limit of 18m.[1]

As some commentators would have it, market forces have taken over from the control once exercised by the major oil companies and thereafter, until the mid-1980s, by OPEC. Yet it would be nonsense to conclude that politics in the Middle East generally and the Gulf region in particular have ceased to be of concern to either producers or consumers. On the contrary, oil was arguably the single most important factor at stake in the 1990–91 Gulf war. Even before that, in 1987, the United States took on the task of protecting Kuwaiti oil tankers at risk from Iranian attack in retaliation for Iraqi strikes on Iranian shipping in the Gulf. The Soviet Union, various European countries and Japan all made contributions, either directly or indirectly, to the protection of freedom of navigation during the Iran–Iraq war. Meanwhile, as a direct consequence of the war, Iraqi and Iranian oil production was substantially damaged or curtailed. If other producers were able to make up the difference and thereby serve the interests of consumers, that too must count as a consequence of regional politics and international involvement.

External intervention to preserve freedom of passage in the Gulf during 'the tanker war' formed the prelude to the historic developments that followed Iraq's invasion of Kuwait in August 1990. Oil had a great

[1] Daniel Yergin, *The Prize: The Epic Quest for Oil, Money and Power,* Simon and Schuster, 1993, p. 719.

deal to do with the Iraqi decision to go into Kuwait. In the preceding months oil prices were on a downward slide, prompting some OPEC producers to maintain their revenues by pumping more, in Kuwait's case in excess of its OPEC quota. Iraq did not have the spare capacity for the same recourse and consequently the Baghdad government remonstrated with Kuwait, as well as accusing the Kuwaitis of horizontal drilling in the Rumailah field which straddles the Iraq–Kuwait border. Iraq also wanted forgiveness of the debts it had incurred to Saudi Arabia and Kuwait during its war with Iran. Saudi Arabia conceded, but Kuwait did not. Against this background, Iraq acted on a claim which it has long harboured to sovereignty over Kuwait, and its tanks rolled into the emirate on the morning of 2 August 1990.

Members of the international community were quick to condemn the Iraqi invasion and sanctions were imposed, endorsed by the United Nations, that prevented Baghdad from exporting its own or Kuwaiti oil or importing any material other than food and medicine. Whether or not the United States had connived at the Iraqi decision to lay claim to Kuwait, it seems that it failed to convey to Baghdad the consequences which would ensue. During autumn 1990 a massive force of US and allied military might was assembled in Saudi Arabia and neighbouring states, initially, in the phase called 'Desert Shield', to protect the kingdom (and particularly its oilfields in the Eastern Province) from an Iraqi advance, and thereafter, in what became 'Desert Storm', to drive the Iraqi forces from Kuwait. In the war which lasted from 17 January to 28 February 1991 Kuwait was duly liberated and Iraq itself was systematically bombed by coalition planes to destroy its infrastructure and communications system. By the finish it was thought that it would take years for Iraq to recover and rebuild and, in Washington at least, the perception was that the Iraqi regime could not survive the humiliation.

Nonetheless, allied forces did nothing to prevent Iraqi troops from crushing a rebellion which erupted in southern Iraq, and they intervened only later to rehabilitate Kurdish refugees who fled to the Turkish border after a failed uprising in the north. In the midst of the chaos 'no-fly zones' were imposed in north and south Iraq, policed by US and allied air forces operating from Turkey and Saudi Arabia respectively. These have endured to 1998. So too has the sanctions regime originally imposed in 1990. On 3 April 1991 the UN Security Council passed Resolution 687, which embodied the terms of a permanent ceasefire and required Iraq to reveal and dismantle its chemical, nuclear and biological weapons programmes and its long-range missile capabilities. A UN Special Commission (UNSCOM) was set up to oversee fulfilment of this task along with the International Atomic Energy Agency. As will be discussed subsequently, by mid-1998 UNSCOM was still not satisfied that its work was done and the UN sanctions remained in place accordingly, as stipulated in Resolution 687. Because of the extended duration of the sanctions, a set of additional resolutions has recently permitted Iraq to export some oil and spend the proceeds, under UN supervision, on 'humanitarian' goods.

The net effect, however, has been Iraq's official exclusion from the oil market for most of the 1990s, apart from limited supplies going to Jordan. The consequences of the extended embargo for Iraqi society are discussed in Chapter 6. At the regional and strategic level, meanwhile, the sanctions on Iraq have had broader consequences for energy. When Iraq and, for a time, Kuwait went off the market in 1990, Saudi Arabia adopted its established role of protector of consumer loyalty and virtually doubled production of crude, to around 8m bpd, to make up the shortfall. An initial oil price spike in autumn 1990 proved short-lived. In a sense, therefore, Gulf producers generally could claim that stability was preserved and even a major war in the Gulf need not frighten the oil industry and consumers. The problem, however, is that the uneasy state of affairs in the Gulf which emerged after Desert Storm was predicated on a continuing US presence and set of policies which could not be expected to endure indefinitely.

In addition, having vastly increased its output, Saudi Arabia developed a new policy to match, based on maintaining a commanding market share within OPEC. Consequently, the likelihood that the kingdom would again cut back production to accommodate a return of Iraq to the market has been cast into doubt. Thus, when there has been speculation that Iraq might finally be near to receiving a clean bill of health from UNSCOM, the market has reacted and prices have dropped. Conversely, on the numerous occasions during the 1990s when Iraq has come into confrontation with the United States and allied forces over compliance with the various UN resolutions that still govern its situation, or operation of the no-fly zones, prices have rallied, on the expectations of military action and/or the prolongation of sanctions. From the end of 1997, however, UN provisions for Iraq to sell oil even under sanctions have contributed to the price collapse at the turn of the year.

The United States holds the balance

Commentaries on the commanding role assumed by the United States in the 1990–91 Gulf crisis and war over Kuwait frequently express the view that were it not for the end of the Cold War and the demise of the Soviet Union, it would not have been possible for Washington to take the lead without challenge. In any case, the situation marked a turning point for US power and influence in the region and, when the war was over, even though the Americans and others made haste to remove the forces that had accumulated to fight, the United States led the way in establishing a new security regime. The US posture in the Gulf encompasses a range of relationships and policy instruments. These include economic assistance, principally to the parties directly involved in the Arab–Israeli peace process; the imposition of trade embargoes on states considered a threat to peace (Libya, Iraq, Iran); political alliances or understandings with states across the region (including Turkey, Egypt, Israel, Jordan and the GCC states); military assistance programmes and defence sales to these allies; and a forward US military, air and naval presence in the Gulf (as well as the Mediterranean).

In a document issued by the United States Central Command (CENTCOM), the military command charged with defending US security interests in the Gulf region and the Horn of Africa, the most vital of those interests were defined as follows:

> maintaining the free flow of oil and assuring freedom of navigation and access to commercial markets are among our nation's vital interests. Other vital interests are safeguarding the security of American citizens and property abroad and promoting the security of regional friends in the context of a comprehensive Middle East peace.[2]

The United States consumes around a quarter of the average daily world oil output; more than half of US oil consumption is imported. OPEC's

[2] US Central Command, *Strategic Plan* 1995–97, p. 5.

share of US crude oil imports in the first quarter of 1998 amounted to 46 per cent of the total 8.1m bpd imported. Venezuela and Saudi Arabia were the principal suppliers of the US market, with imports from these two countries averaging 1.324m and 1.332m bpd respectively.[3] Aside from this considerable reliance on Saudi crude, however, US imports from other Gulf producers averaged only about 300,000 bpd in early 1998. All imports from Iran ceased when new regulations came into effect in 1995.

The approach to the Middle East espoused by US President Bill Clinton in his first administration was designed to integrate US policy in the Arab–Israeli sector with US strategy in the Gulf. Two of the most significant threats to regional stability in both areas, according to the Clinton administration, were the proliferation of weapons of mass destruction and terrorism. Israel was seen to be in danger from both these threats, and certain states, dubbed by the administration as 'rogue states', have been singled out as the principal enemies of regional peace, because of their support for terrorism and their ambitions to acquire nuclear as well as chemical and biological weapons. (The US list of 'rogue states' worldwide includes Iraq, Iran, Libya, North Korea, Cuba and Sudan.)

With respect to Iraq and Iran, the strategy of so-called 'dual containment' was initially articulated in 1993 by then National Security Council member Martin Indyk, who said the concept

> derives from an assessment that the current Iraqi and Iranian regimes are both hostile to American interests in the region. Accordingly, we do not accept the argument that we should continue the old balance of power game .... we reject it because we do not need it. ... The coalition that fought Saddam remains together. As long as we are able to maintain our military presence in the region; as long as we succeed in restricting the military ambitions of both Iraq and Iran; and as long as we can rely on our regional allies – Egypt, Israel, Saudi Arabia and the GCC, and Turkey – to preserve the balance of power in our favour in the wider Middle East region, we will have the means to counter both the Iraqi and the Iranian regimes.[4]

[3] 'US Imports of OPEC Crude Increase In January–April 1998,' *Middle East Economic Survey (MEES)*, 6 July 1998.
[4] *Middle East International,* No. 452, 11 June 1993, pp. 3–4.

During the course of the 1990s, the administration, pressured by Congress, adopted an increasingly uncompromising tone on Iran and force was used periodically to keep Iraq 'in its box', to use the phrase coined by US Secretary of State Madeleine Albright.

However, by 1998 it was apparent that regional dynamics were overtaking US policy. The Arab–Israeli peace process ran into trouble and eventually ground to a halt in the eighteen months following the election of the Likud-led government of Prime Minister Binyamin Netanyahu in Israel in May 1996. The close relations developed by the Clinton administration with Israel's previous Labour government were put under strain, but Washington's apparent failure to force Netanyahu to respond to its bidding on the peace process became a source of frustration and anxiety for America's Arab allies in the region, as well as for the Europeans. Dissident opposition to the US military presence in Saudi Arabia seemingly lay behind a bomb attack on a US mission in Riyadh in 1995 and another on the Al Khobar Towers housing US forces in Dhahran in June 1996. The latter prompted a relocation of the forces to a remote spot in the desert south of Riyadh.

In Iraq, US efforts to help plot the overthrow of Saddam Hussein collapsed ignominiously in 1996,[5] as is discussed in Chapter 6 by Raad Alkadiri. Nearly two years later, when the United States tried to mount a military strike to deal a potentially fatal blow to the Iraqi regime for its defiance of UN weapons inspectors in February 1998, the only ally to offer unqualified support was Britain. Washington's Arab allies expressed concern, if not outright opposition to another round of bombings which might result in chaos or the strengthening of Saddam Hussein's hand.

Meanwhile, developments in Iran produced a landslide victory for Mohammed Khatami in the May 1997 presidential elections. His emphasis on the need to develop civil society and uphold the rule of law on the domestic front, together with his adoption of a new tone towards the West, signalled that significant changes could be on the way in the Islamic Republic. This turn of events gave credence to the claims of some senior figures in the United States that its policy on Iran was ripe for reappraisal. In any case, the US secondary boycott legislation embodied in the Iran and Libya Sanctions Act (ILSA) of 1996, designed to prevent

[5] Patrick Cockburn, 'CIA Botched Saddam Overthrow', *Independent*, 11 April 1997.

foreign companies from investing in the Iranian energy sector, had already sparked such a furious row with America's European allies that the administration decided to issue a waiver for Total, Gazprom and Petronas in 1998, notwithstanding their blatant challenge to its provisions.

All change

In Chapter 4 on US policy, Gary Sick concludes with the time-worn maxim that the only certainty is change. From the vantage point of mid-1998, it is fair to conclude that the 'moment' of US hegemony in the Middle East, brought about by a particular convergence of events at the beginning of the decade, has already peaked. US actions to combat terrorism, notably the missile assaults on sites in Afghanistan and Sudan in August, may imply Washington's defiance of its self-declared Islamist enemies such as Osama Bin Laden, but public opinion across the Arab and Muslim world generally has turned increasingly hostile to the United States, presaging a new level of resistance to its influence. For those in the Gulf this development came on top of concern about the impact of an oil price collapse in mid-1998.

This phenomenon was so serious that Saudi Arabia and Venezuela decided to forget their erstwhile differences and reach agreement with Mexico, a non-OPEC producer, to cut back oil production. This was the first of a series of moves heralded by Saudi Arabia as the beginning of a new world order of energy producers. In any case, the price slump accelerated the pressures on Gulf producers to find new ways to structure their economies as a whole. This is necessarily a long-term objective, and in the meantime there could well be another major political crisis in the Middle East, possibly centred on Iraq, but almost inevitably involving Arab–Israeli relations.

In sum, as of this writing in summer 1998, the chances seem high that the Gulf region could be about to witness a new period of upheaval and uncertainty. This volume provides an overview of the issues at stake and the factors most likely to drive the changes under way.

2 Global setting and trends for oil and gas

John V. Mitchell

This chapter provides a context within which to view the prospects for Gulf energy producers in general and the United Arab Emirates in particular. As a preliminary to a detailed discussion of the issues affecting demand, supply and prices, some remarks about methodology are in order.

Methods and madness

When making predictions, considerable caution is required. Especially in the energy sector the record is very mixed. Numbers are essential, because without them it is difficult to tell what is important and what is trivial. However, any given number may be inaccurate, so one has to have some idea of how the numbers are connected to one another – in other words, one has to have some kind of model. The invention of algebra by Arab mathematicians so many centuries ago is a case in point. Whether a model is good or useful depends on two quite separate criteria:

(1) The model must be logical and consistent within itself. To guarantee this it is prudent to use models which are based on good disciplines – in this case economics – which, like algebra, have been developed and tested over many years.
(2) The chosen model should correspond to the real world as we observe it. This requirement is much harder to meet, since the real world is complicated and difficult for humans to understand even before they try to predict the future. There are many economists who have made great mistakes about energy matters because their

> logical models did not fit the real world.[1] Yet many people with great understanding of the real world have made serious mistakes about energy because they did not fit their observations into a disciplined model.[2]

This chapter is intended to determine in which respects a given model measures up to these criteria and which of the available numbers inspire confidence. This will highlight where there are risks because either there is no sensible model or there are no sensible numbers.[3]

To focus the discussion, a comparison will be made between two cases or sets of projections, illustrating what makes them different in terms of 'models' and of 'numbers'. The purpose of the exercise is both to explore the usefulness of the tools for making projections, and to determine in what ways alternative forecasts may be 'optimistic' or 'pessimistic' for Gulf producers.

Examination of Case 1 (HOPEC)

The first case represents more or less the conventional wisdom of the energy establishments of the importing countries. The numbers for 1997 and 1998 are based closely on the International Energy Agency (IEA) monthly oil market reports. The numbers for 2010 and 2015 are based on the 'reference case' of the US Department of Energy Information Agency (EIA) *International Energy Outlook* published in April 1997.[4] The latter is a very transparent forecast: the assumptions are set out clearly and are very conventional as regards economic growth and so on. There is also a

[1] The use of the ' Hotelling Principle' in the early 1980s to forecast ever rising oil prices is a good example of this.

[2] The oil industry managers who in the early 1980s refused to believe that tripling the oil price would reduce the demand for oil are an example of this.

[3] For anyone who is interested, in my own analysis of the oil industry I follow the ideas of the American economist Michael Porter, who has tried to analyse, for industry in general and nations in general, the dynamics of competition, technical change and the interactions between different industries.

[4] *The International Energy Outlook 1997,* Energy Information Agency, DOE/EIA 0484 (97), US Department of Energy, April 1997 (EIA 1997). The 1998 *International Energy Outlook,* published in May 1998, provides a similar view. This source is the only publicly available long-term energy forecast by a major public body. The IEA has not published a long-term forecast for 1997, and the European Commission's last relevant forecast was published in 1995.

range of scenarios to illustrate the effects of different assumptions about key variables such as economic growth and energy intensity. There is no other set of projections publicly available which is so transparent and well documented. Both the projections and the computer model itself can be downloaded from the Internet.[5] For present purposes, this case is acceptable as a 'model',[6] though that should not be taken as a professional evaluation or endorsement. It is a convenient and respectable peg on which to hang comments about the relation of this or any similar model to the 'facts' or what the facts of the long-term future may be. The assumptions and inferences to be drawn are more open to debate. Figure 2.1 shows the EIA reference case. Here it is termed the high OPEC or 'HOPEC' case. In it OPEC oil production rises to 53 per cent of world oil production by 2015 with 42 per cent derived from the Gulf region.

The key issues in projections like these may be divided into two broad categories: demand and supply.

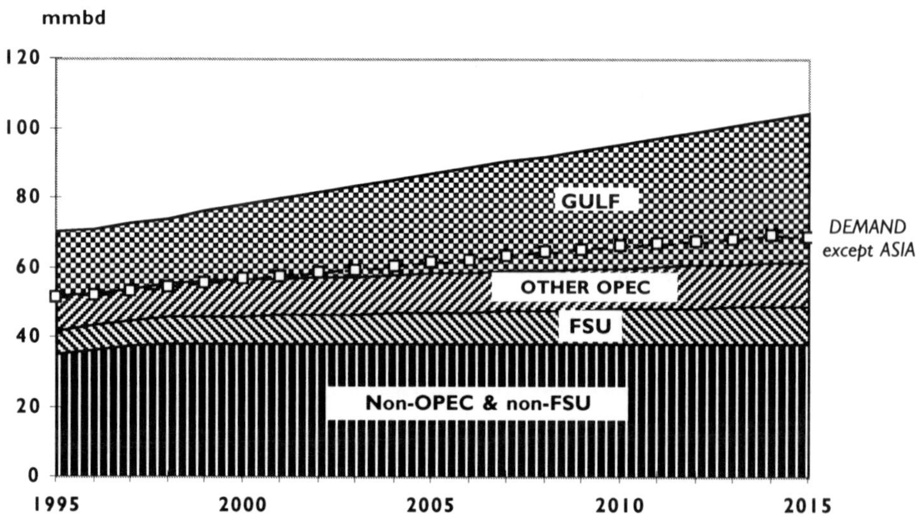

Figure 2.1: Case 1 projection (HOPEC)

[5] http://www.eia.doe.gov.

[6] Several computer models are used to prepare the forecast. I have no reason to doubt their integrity or economic logic.

Demand issues

Economic growth

The drivers of economic growth are technical changes of all kinds, which can be expected to continue in the future. Economic gains will derive from the combination of new information and communications technology on the one hand, with the liberalization of markets and the privatization of enterprises on the other: cheap and easy communications make competitive markets fairer and more efficient.

The EIA's projection for economic growth was about 5 per cent higher for 1997 than it was for 1996, reflecting a generally more optimistic view about long-term growth rates in the United States and Asia. It is to be expected that the 1998 projections will be reduced to reflect the loss of growth immediately after the 1997–8 Asian financial crisis.[7]

The EIA projection assumes that world economic growth over the twenty years 1995–2015 will average 3.1 per cent, a rate one-tenth higher than the average over the twenty-five years 1970–95. The past twenty-five years included three macroeconomic shocks: the oil and inflation shock of 1973, the oil and deflation shock of 1979–80, and the 'Black Monday' financial shock of 1987. Each of these was followed by a period of deep and long recession in the industrialized countries. Such shocks move the baseline for subsequent growth. OECD energy consumption ran below its 1979 level for seven lean years from 1980 to 1986. Another macroeconomic shock – for whatever reason – would shift the projected energy demand curves to the right, reduce the projection for any future date like 2020 or 2015, and reduce the average rate of growth over the same period. Clearly, it is important to understand the simple arithmetic. An expectation of smooth growth from 1996 or 1997 levels probably overlooks the effect of future recessions.

[7] The EIA projection for May 1998 marked down Asian oil demand forecasts by 3–500,000 bpd for 1998–9, but then projected a 'bounceback' so that the estimate for 2015 was similar to that in the 1997 forecast. A more conservative assumption might be that Asian oil demand will not grow for three years and that the former rate of growth would then resume. The effect of this would be that during those three years Middle East exporters would see no increase in the demand for their oil, since the growth in demand in the rest of the world would be met completely by increases in production outside the Middle East.

A more optimistic view of the future (like the EIA reference case) implies either that there will be no more macroeconomic shocks, or that the industrial countries will follow the United States, in Alan Greenspan's words, in some 'new paradigm' of higher than normal growth with no serious recessions. From the perspective of history, all such hopeful projections are 'optimistic' from the point of view of the producers of oil. This comment applies particularly to Asia, which in most projections shows the most rapid and sustained growth in the economy and therefore in energy demand. The EIA reference case projects an *average* of 6 per cent economic growth for Asia (including over 7 per cent for China), over twenty years, a rate slightly below recent history but high by the standards of rapidly growing areas over long periods hitherto. Figure 2.1 also shows that the increase in oil demand which is attributable to Asia is roughly equal to the increase attributed to Gulf oil production. The future of West Asia as an energy-exporting region and East and South Asia as an importing region will become more and more a shared future. The simple arithmetic of projecting high rates of growth in substantial markets over a long period means that uncertainty in the final years is very high.

Recent events in the Asian financial markets, in the opinion of this author, are likely to create a pause in growth rather than a permanent reduction of the potential growth rate (which in any case was projected to fall slightly). But when there is a pause in high growth rates, the arithmetic of compound interest will have an effect on the forecast for a particular year.[8] If East Asia grows from 1998 to 2000 at 2 per cent instead of its supposed potential future rate of 7 per cent, and then returns to 7 per cent growth from 2000 to 2020, the projection for the specific year 2020 would need to be reduced by about 14 per cent. Moreover, the need for that adjustment would soon become obvious – within three years – while the question of whether or not to adjust projections of long-term potential growth rates might remain open for a longer period. So it should soon be possible to make at least a first decision about the effect of the Asian financial crisis on projections for 2020.[9]

[8] This was a large part of the explanation of mistakes made in forecasting in the 1980s for oil demand in 1990: not that the growth *rate* was wrong but that the starting point was delayed by recession.

[9] See Note 7 above. With no oil demand growth for three years the reduction in the figure previously projected for 2020 would be nearer 20 per cent. Such are the cumulative effects of errors in high growth rates.

Relating energy demand and economic growth

The next stage to consider in the forecasting process is the relationship between the demand for energy and projected economic growth. There are several elements in this relationship: the development of new technology to make manufacturing and transportation more efficient in all respects – including the use of energy; the diffusion of that technology through the economic system by competition and liberalization of markets; and government policies which for environmental reasons aim to curb the unwanted pollution that occurs when fuels are used, and therefore enforce the adoption of more efficient energy consumption technologies, the use of less energy and perhaps even some loss of economic growth.

In the EIA reference case, world economic growth of an average 3.1 per cent over twenty years is projected to generate energy consumption growth of 2.2 per cent – nearly one-third lower than economic growth. Many forecasters would probably accept '3 per cent GDP growth and 2 per cent energy growth' as a benchmark against which to work because that was the story of the 1980s. However, over the period 1970–95 as a whole, and during the 1990s, energy demand worldwide has grown at over 80 per cent of the growth rate of GDP. Did the 1980s reflect the high prices of 1979–80 and do the 1990s reflect the lower energy prices since 1986? If so, how should we project the relationship in a world without dramatic increases in energy prices over decades starting from the oil price collapse of 1998? In fact, it seems reasonable to conclude that a 3 per cent GDP/2 per cent energy projection actually forecasts that the relationship will change in a way which is 'pessimistic' for energy producers (implying more energy efficiency) but 'optimistic' for those concerned with climate protection and pollution (looking for lower absolute levels of fuel use).

Any reputable forecaster will admit, as the US EIA does,[10] that the relation between energy demand and economic growth is a most difficult topic. There are no top-down formulas that work. The bottom-up attempts by experts such as Lee Schipper are very useful in explaining the past and warning about the dangers of simple trends, but this method requires

[10] See EIA, *International Energy Outlook 1997*, p. 19.

forecasting of huge numbers of unpredictable items – technical change in industry, changing consumer tastes, the distribution of income and so on. Instinct suggests, to this author at least, that it is fair to project a changing relationship for the developing countries, but more doubtful for the present industrialized countries.

Broadly, the EIA assumption is that in *developing countries* by 2015 energy consumption will grow at about 60 per cent of the rate of growth of GDP – the same sort of relationship that prevailed in industrialized countries in the 1980s and 1990s, and a significant change from recent history when energy consumption and GDP have grown at the same rate on average over long periods. Some change of this kind is reasonable to expect, as the larger developing countries become more industrialized and open their economies to market mechanisms and liberalize prices.

Over the long run, there is evidence that energy prices influence the relationship as economists would expect: higher prices lead to more efficient use of energy. There are large areas of the developing world – the former Soviet Union (FSU), China, Brazil, India and many energy-exporting countries – where energy prices to consumers have been kept low by price controls or subsidies, or where the choice of energy-using equipment has been made by central planners rather than market forces. Not surprisingly, most of these countries extract lower GDP from their use of energy than market economies. The same 'marketization' and liberalization which will generate continuing high rates of economic growth will also raise domestic energy prices in developing countries and emerging market economies where they have been subsidized or held down by price controls. This will be the reverse of the experience in European liberalization in the 1980s, where prices started from international levels but had been marked up by distributors' and retailers' margins which were reduced by competition. The situation will be more like the end of price controls in North America in 1979–81. As in North America, so in the newly industrializing economies and emerging markets, higher prices, more competition and liberalized markets will provide incentives for introducing energy-efficient technologies and better-managed, energy-intensive operations, in the same rigorous way as is happening in the industrial countries. A gap will open between GDP and energy growth rates in the developing countries. The relationship between economic growth and energy demand in the newly industrializing countries can be expected to resemble the relationship already achieved in the industrialized market economies.

Nevertheless, there is great uncertainty about how far the changes will go, how long they will take and what their effect will be on developing economies which are rapidly changing. Figure 2.2 illustrates this by showing a range of point forecasts of energy demand for 2010 for China which were published in 1996.[11]

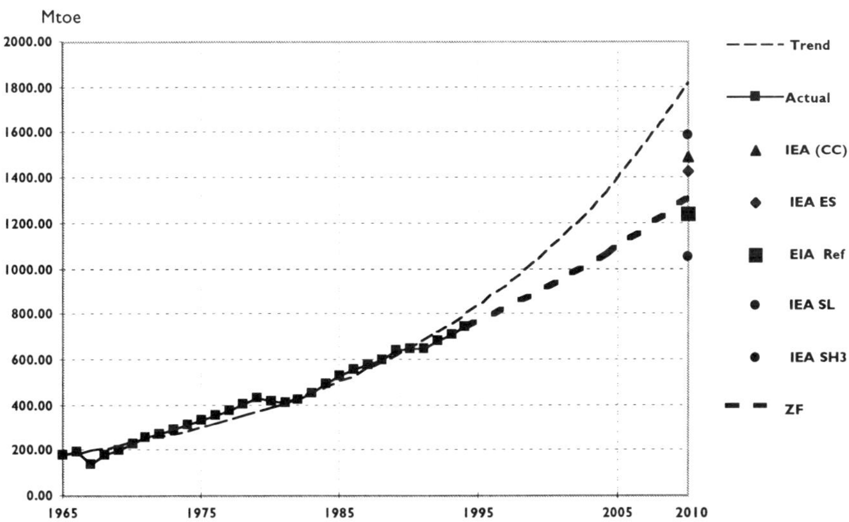

Figure 2.2: China: primary energy consumption

All the projections for 2010 are lower than the statistical trend line (the thin broken line) – reflecting a common assumption about a widening gap between GDP growth and energy growth. The heavier broken line (ZF) is a 1995 State Planning Commission forecast. The point is that the range between reasonable high and low forecasts for 2010 is three times the current level of Chinese energy consumption.

In the 1997 EIA projections for the currently industrialized countries, the rate of growth in energy demand is expected to fall to less than 50 per

[11] The EIA references are to the 1996 EIA *International Energy Outlook*. Further and more recent comparisons are given in the *International Energy Outlook* for 1997 and 1998.

cent of GDP growth. Since the reference case in these projections (and many similar projections) does not assume dramatic increases in average energy prices, such changes depend on continuing technical change – 'demand-side' management of energy with better information and control systems – and/or a degree of regulation to limit consumer choice (for example by 'CAFE' (Corporate average fuel efficiency)-style standards for vehicles, efficiency standards for residential and commercial appliances, and voluntary programmes for industry: these could be imposed if serious climate change policies develop). The effect of assuming such changes in the reference case is to reduce the severity of the demand restraint measures which would be needed to achieve a more environmentally tolerable outcome.

Dividing the energy market among fuels

The next step in the process of demand projection is the allocation of energy demand (and therefore supply) among different fuels. For the heat and electricity markets, this has to be done regionally, since hydro-electricity is concentrated in some countries and nuclear power (given current political constraints) in others (a few have both). Because of transportation costs, the largest gas and coal shares of energy markets are also in those countries which have developed gas and coal resources.

For the transport market, which may currently account for about 60 per cent of world oil consumption,[12] oil is almost the fuel of necessity as well as choice and is likely to remain so for many years in many countries. Conventional wisdom predicts a continuation of trends in potential fuel efficiency for a variety of technical reasons (improved combustion, exhaust management, reduced drag and rolling resistance). Higher fuel prices and regulations to reduce fuel consumption in order to reduce CO_2 emissions will continue to stimulate the exploitation of this potential in the developed countries. The global nature of the automobile industry means that these technical advances will be rapidly transmitted to

[12] IPCC, Second Assessment Report. Chapter II.C.3. Transportation, First Order Draft 1994, cited in World Energy Council, *Global Transport Sector Energy Demand Towards 2020,* London, World Energy Council, 1995, p. 9. The figure is an estimate based on poor data, since the final use of middle distillates is not recorded in many countries' energy statistics. Other sources estimate lower figures: for example N. Wohlgemuth, 'World Transport Energy Demand and Modelling', *Energy Policy,* Oxford, Elsevier Science, Vols 14–15, p. 1110.

developing countries which are open to international trade, and therefore to global competition, between the major automobile and component manufacturers based in the developed countries. It is less certain how economic growth in the developing countries will translate into transport systems and therefore vehicle demand: vehicle ownership levels are very low in many developing countries and particularly in China. Options are still open in these countries to halt or slow down the development of urban congestion problems such as those affecting Bangkok, Shanghai, São Paulo and Djakarta. Urban public transport remains an option – though an expensive one. Many of these 'system choices' will be made, often by default, within the next decade – but until then the twenty- or thirty-year rate of growth of transport fuel demand in many major developing countries remains very uncertain. Most mainstream projections (including those of the EIA) reflect a decline in the share of nuclear power in the electricity sector worldwide as new nuclear plant building is confined to China and South Korea, with other OECD countries failing to add or even replace plants.

There is also general agreement that there will be an increase in the gas share of energy demand in Europe and in the United States. It is the development of Combined Cycle Gas Turbine Generation Technology (CCGT), with its higher efficiency, low capital cost, and relatively small optimum plant size, that has opened the way for aggressive gas penetration into the electricity markets in Europe, where these have been deregulated. In the United States deregulation of both the gas and electricity markets, combined with the impact of gas prices held down by gas-to-gas competition, has strengthened the gas share of the industrial market. These factors are not present in many developing countries, but the trends are in that direction. On the market side electricity sectors are liberalized and privatized, often with the result that previously subsidized or artificially low power prices are raised. At the same time suppliers with undeveloped reserves – in the Gulf, Central Asia, Russia and Indonesia – have hoped to get their reserves into the Asian market during its rapid expansion. The case of Asia is problematic, since increasing market shares for gas require the expansion of LNG supplies and the development of pipeline supplies from Central Asia and east Siberia. Very large investments in new infrastructure will be required. In Europe over the next ten years at least there will be a 'supply push' from exporters such as

Table 2.1: EIA reference case: gas percentage share of energy market

	1995	2015
Industrial countries	21	27
Developing countries and EMEs	21	25

Russia, Norway and Algeria. In Latin America, upstream expansion will provide supplies which will be used in power production.

The EIA reference projection, in Table 2.1, like most mainline projections, foresees gas gaining market share, in both industrial and developed countries. This may understate the potential for gas penetration. To give a simple example from Europe: if gas were to gain the same market share in all European countries as it now has in Britain or the United States, gas demand within Europe would be about 3m bpd oil equivalent higher than is assumed in the EIA scenario: some of this would be at the expense of oil, particularly in the space heating and light industrial sector; some would be at the expense of *potential* oil markets as the nuclear share of the electricity market falls.

There is a multiple hazard in these forecasts: it is necessary first to forecast the electricity market by region, and then the fuel shares within that market. In the United States the large coal industry has met the threat of competition from gas by reducing costs (helped by the deregulation of the transport industry, leading to reductions in costs for low-sulphur Western coal – which the industry required to meet tighter restrictions on sulphur emissions). In Russia, replacing the old 'one price for Russia' formula in the coal industry led to an increase in transportation costs which worked to the disadvantage of coal. For China, there are contrary forces at work. There is a need to reduce urban pollution by switching to cleaner coal as well as a need to reduce the demand for transport infrastructure by upgrading coal at the mine. Economic pricing for transport will have mixed effects, which will vary from region to region. This will probably open potential markets for gas in some parts of China if investments in infrastructure are made. The economics of such investments, however, depend on views about future electricity prices in an economy less centrally planned than today and in which major sectors will be privatized: it is impossible to know the outcome of this.

Summary on the demand side

All one can say is that the kinds of figures involved at a global level reflect – with inevitable inaccuracy – trends whose direction can be predicted with reasonable confidence. Whether the projections can be made more accurate is doubtful, but it makes sense to concentrate on the following factors:

- the rates of economic growth in different parts of the world;
- the development of the electricity markets and the rate of penetration of gas through CCGT over the next twenty years, and of renewable energy sources in the longer term;
- those areas where direct-injection combustion engines and improved vehicle designs take the biggest share of future vehicle fleets.

Government policies on pollution, urban congestion and the environment are also important for the very long term but, in the view of this author, for some years it will be difficult to quantify their effects. Strong policies are likely to be adopted slowly, and their effect will depend on the modification and renewal of the capital stock of energy-using capital equipment over several decades.

Oil supply

Sources of supply: the conventional wisdom

Figure 2.1, which shows demand, also gives a very conventional breakdown of future supply. The three key features are:

- a growing call on the Gulf as the incremental supplier, so that Gulf production in this projection would reach 35m bpd by 2010 and 43m bpd by 2015;
- steady growth in supplies from OPEC outside the Gulf and from the FSU;
- a flattening out of growth in the rest of the world, the 'non-OPEC and non-FSU' (neither OPEC nor the FSU) at around 38m bpd soon

> after the year 2000. This is about 5m bpd higher for 2015 than the EIA's projection in 1996 and reflects widespread belief that technical progress will continue to reduce costs and risk of discovery of new reserves, open the way to economic development of small reservoirs, and increase recovery rates from developed fields.

This 'flat' projection assumes a fundamental change in the production trend for what is a very large part of the world. For the period 1965–96, the 'rest of the world' increased production steadily every year by between 600,000 bpd and 700,000 bpd, falling only in two years (and growing more rapidly in some recent years). This is shown in Figure 2.3.

There are two processes at work: one is the assumption that at some time early in the twenty-first century the limited size of then known oil reserves will slow down and eventually reverse the growth of oil production. The second assumption is that this effect will occur first outside OPEC and the FSU – precisely in those areas where production had grown most steadily. For the countries in this category as a whole, the ratio of currently proven reserves to current production is lower than in most OPEC countries.

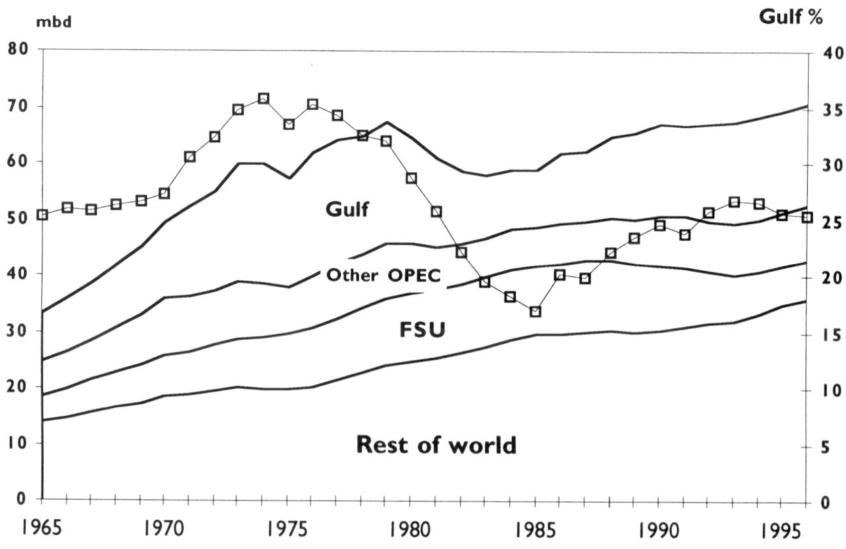

Figure 2.3: Oil production 1965–96

The debate about depletion and the limits to growth is ongoing, though the proponents of the limits theory have lost credibility because their predictions of reserve constraint have so often been proved wrong. There is also the problem of knowing the reliability of the figures used for proven reserves, before making any projection of future reserves.[13] For the geologists and engineers, the key questions are:

- the rate of depletion of known reserves;
- the falling size of new discoveries;
- the extent to which increased recovery rates merely advance production without increasing total recovery.

There is, however, a serious debate between petroleum engineers and geologists such as C.J. Campbell and J.M. Laherrère,[14] and economists such as Professor Adelman of MIT and many others (including this author). Robert Mabro has reviewed the issues[15] and they were discussed at a workshop at the University of Reading, England, in 1997.[16] For the economists, the key questions are:

- whether technical progress will continue upstream;
- whether market forces will be allowed to work – so that market prices reach the producer without being depleted by royalties and taxes;
- whether fears of environmental damage will lead governments to limit exploration and development. (In the United States potential recoverable reserves are shut off by state and federal government concerns about damage to ecosystems in wetland and sensitive areas. Except for Alaska, Texas and Louisiana, every maritime state in the United States has banned further leasing of acreage for offshore exploration.)

[13] For a review of the statistical and reporting issues, see Gilbert Jenkins, 'World Oil Reserves Reporting 1994–96: Political, Economic and Subjective Influences', in *The OPEC Review,* Vienna, Organization of Petroleum Exporting Countries, June 1997.

[14] C.J. Campbell and J.M. Laherrère, *The World's Oil Supply 1930-2050*, Geneva, Petro Consultants, 1997. See also Craig Bond Matfield, 'How Long Can Oil Supply Grow?', Hubbert Center Newsletter 97/4, M. King Hubbert Center for Petroleum Supply Studies, Colorado School of Mines, Golden, CO, 1997.

[15] Robert Mabro, 'The World's Oil Supply, 1930–2050', a review article, *The Journal of Energy Literature,* Vol. II, No. 1, Oxford Institute of Energy Studies, 1966.

[16] See Bruce Sellwood and Max Coleman, 'Future Prospects for Conventional Supplies', in *Petroleum Review,* London, Institute of Petroleum, December 1997.

These arguments have been set out by this author elsewhere,[17] by Professor Peter Odell[18] and more technically in a recent World Bank paper by G.C. Watkins and Shane Streifel, who attempt to estimate an oil supply function econometrically.[19]

An alternative projection: Case 2 (LOPEC)

By comparing the 'HOPEC' case explored above with an alternative projection, the inherent characteristics of the former case become manifest. The conclusion drawn here is that the HOPEC case is 'optimistic' from the point of view of the demand on the Gulf, and 'pessimistic' about the prospects of sustaining growth in supplies from the rest of the world. This is most easily demonstrated by showing what might be the case if production from the 'rest of the world' ('non-OPEC and non-FSU') did not flatten off but maintained its thirty-five-year trend of steady growth. This alternative is shown in Figure 2.4.

This simple projection of unchanged growth in the 'rest of the world' would make a difference of 9m bpd by 2015. With all other factors unchanged, this would be at the expense of Gulf production, which would on these numbers be around 33 per cent of world production, instead of the 41 per cent which is shown in the HOPEC case for the same level of demand and production in other regions.

The spread of new technology is not limited to the non-OPEC countries, but has been less urgent for producers limited by quotas. If (as is discussed below) there is no upward trend in oil prices, many high-reserve exporting countries will have a strong incentive to increase revenue by reducing the cost of production and the capital required to sustain it. This will increase their long-term production potential in the same way as is happening elsewhere: competition between suppliers will increase within OPEC as well as outside it.

[17] Chapter 2, 'The Oil Machine', in John V. Mitchell, with Peter Beck and Michael Grubb, *The New Geopolitics of Energy,* Royal Institute of International Affairs, London, 1996.
[18] Most recently in Peter R. Odell, 'Oil Reserves: much more than meets the eye', *Petroleum Economist,* London, November 1997.
[19] G.C. Watkins and Shane Streifel, *World Crude Oil,* Resource Policy Research Paper 1756, Washington, DC, World Bank, April 1997.

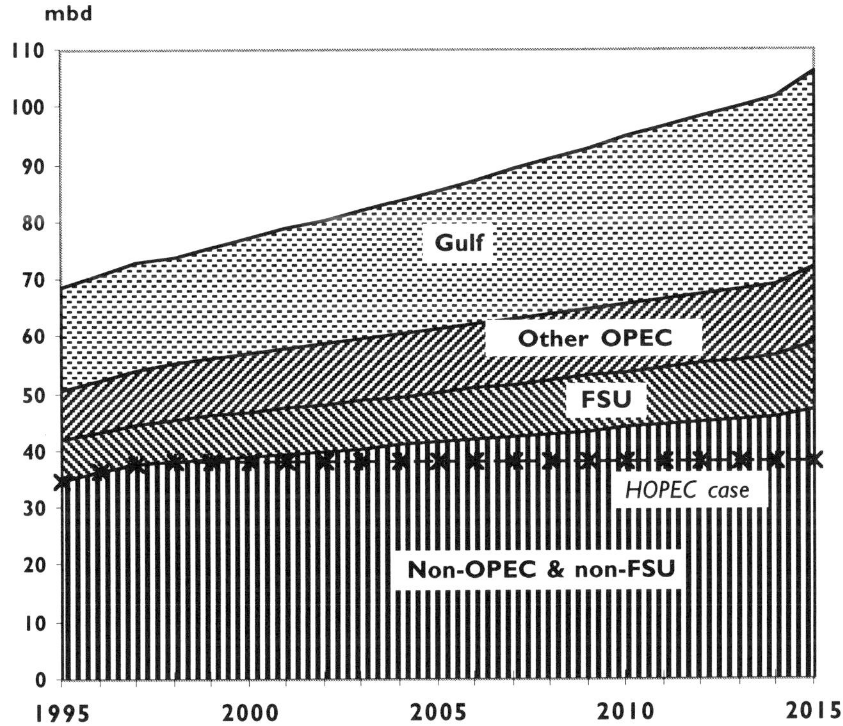

Figure 2.4: Case 2 projection (LOPEC)

Price

Long term (2010–20)

The 1997 EIA projections use a barrel price of $21 in 1995 dollars (about $22 in 1997 dollars) for 2015, with scenarios for $7 higher or lower cases. The methodology of connecting such numbers to the supply and demand figures is inevitably not clear. The connections cannot be quantified very well – there is no algebra! Few people make a good job of forecasting oil prices, even of explaining them. Economic models do not take enough account of political factors. Political analysts do not take enough account of economic constraints. There are conspiracy theorists at each end of the spectrum. Economists expect prices to move down to long-run marginal costs, including the value of the oil if left in the ground.[20] When prices fail to fall to this level economists suspect a conspiracy to prevent competition. Political analysts have looked at the supposed market power of OPEC (in the 1970s) and argued that only a conspiracy of importing governments has prevented prices from being much higher. Attempts to 'model' OPEC behaviour have not been very successful, though there is something to be learnt from such exercises.

With these warnings in mind, it is nonetheless useful to try to connect some of the major factors affecting prices in the long term. These may be summarized as follows:

- the availability and price of oil from outside OPEC;
- the availability and price of natural gas, competing with oil in several markets, sectors and regions;
- the availability and price of oil processed to crude oil equivalent or oil products from heavy oils in the Orinoco basin (Venezuela) and western Canada;
- the availability and price of oil produced from coal (as in South Africa);
- the investment and production policies of the Gulf producers. Some of the Gulf producers have the potential to develop significant volumes of new capacity at low cost, and to reduce production from low-cost developments without suffering immediate economic damage (because of accumulated financial reserves). This gives those producers market power in the medium term (including the medium term of the distant future, within the lead times of developing alternatives), but their market power is constrained by the other factors itemized.

Figure 2.5 is devised to illustrate the effect of these constraints on variations around the level of 105m bpd world oil production projected by the EIA for 2015 in their reference case.

[20] This is not an argument for expecting prices to be $1 per barrel. Few economists would argue that prices could be sustained at variable Middle East production costs. Over the long run prices should recover the cost of development and exploration and the value of oil in the ground, which is calculated by resource economists as the (discounted) value calculated from some future price of alternatives such as heavy oils and nuclear or solar power.

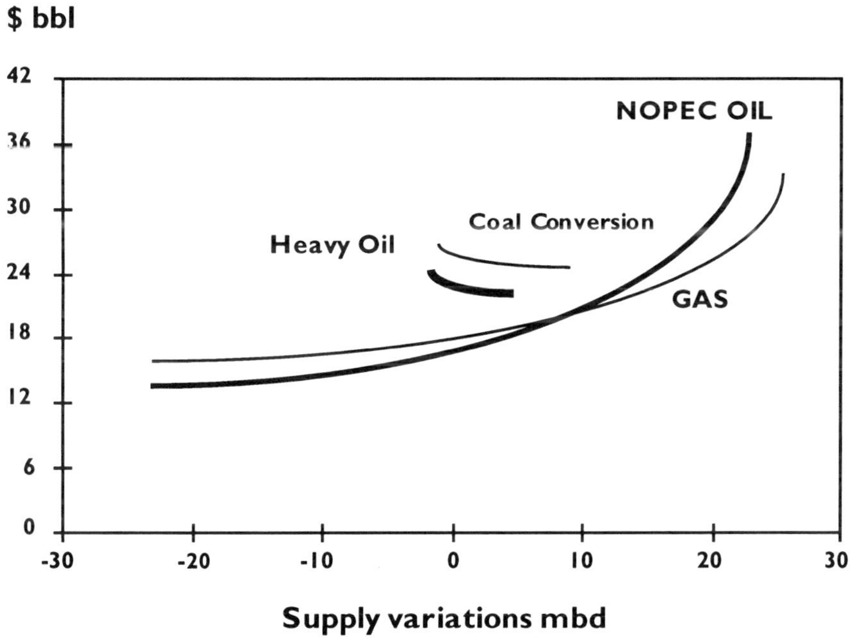

Figure 2.5: Constraints on oil prices, 2015

Pursuing this line of inquiry, certain key ideas emerge – which could be refined by further study and other expertise. First, non-OPEC oil production would be driven down – perhaps by 20m bpd by 2015, by the poor economics of replacing depleted reserves if oil remained at $12 per barrel (bbl). However, non-OPEC production could sustain its historic rate of growth if oil prices were lifted to $24 bbl. The growth would add 10m bpd by 2015.[21] Second, natural gas penetration in Asia would expand substantially if oil prices were sustained over $22 bbl. Against such prices the development of major new pipelines from Russia and Iran, and new LNG projects from the Middle East, could make economic as well as geopolitical sense. Because of the low current penetration of natural gas in Asian markets, and the overhang of known but undeveloped gas reserves, the potential could be large, say 20m bpd by 2015.

[21] In looking ahead, the FSU is assumed to behave under market forces in the same way as the rest of non-OPEC.

A third area for further inquiry is fuel conversion costs. The cost of heavy oil and coal conversion is already coming down, because raw material production costs are being reduced, but the conversion costs can also be expected to come down. Incremental improvements will be apparent as the use of technology and construction processes benefit from the 'experience curve' – at present just beginning in Venezuela. Taking a modest view of three plants being commissioned on average every year over ten years, the effect by 2015 might be an additional 5m bpd. For the even longer term, the effects will grow steadily as the operating base enlarges: unlike offshore oil developments, heavy oil conversion projects (like LNG projects) have a long plateau life, as they are sized around the high capital cost of the conversion facilities. Conversely, a sustained fall to $12 might wipe out the economics of continuing construction and operation even of the present projects.

Finally, the policies of the Gulf producers are critical to the situation at any particular date. Weak investment by Gulf producers will limit availability and drive up prices until investment grows in the supply of alternative sources and competing fuels; weak control of production will have the opposite effect. It is most likely that decisions about capacity expansion and production will in practice be taken mainly for short- and medium-run reasons (discussed below) since these literally have priority over the long term.

Figure 2.5 represents the author's own depiction of the interplay of factors affecting prices in the long term. One should not read anything into the particular points or intersections. To summarize, the general line of argument suggested here runs as follows:

- At around 105m bpd oil supply, and around 1,445 TCF natural gas output, gas supply may be more sensitive than non-OPEC oil supply to lower prices (mainly because these levels assume a substantial share of long-distance pipeline gas and LNG).
- However, over a threshold of around $4/mbtu, the very large Middle East and Russian long distance gas has more potential for expansion than non-OPEC oil.
- Heavy oil and coal conversion benefit from an experience curve – like LNG, once the industry develops, the potential supply could be very large.

- The implication is indeed that non-OPEC oil, even if the reserves were available, would have limited scope for expansion at prices in the low twenties.
- OPEC oil, by contrast, would be in a different position, with a large reserve and lower cost base.
- The picture is thus of an oil market where prices in the long term are constrained by close substitutes, but OPEC, eventually, will gain market share at the expense of non-OPEC.

The general message here is that over the long term OPEC revenues are likely to expand as a result of increasing volume rather than of increasing price.

Medium term

In the medium term (3–5 years) the constraints of long-term gas and heavy oil competition are present only through their effect on expectations. The important variables are:

- economic growth affecting demand. For most current forecasts, this is a downside risk. Recession in the United States (of which there is at present no sign), if it were to occur, could combine with the effects of the Asian financial crisis to reduce expected demand around the year 2000 by 5 per cent or more;
- Gulf producers' investment policies, especially in Saudi Arabia;
- the return of Iraq to the market;
- the evolution of quotas in the event that capacity expands faster than demand.

Broadly speaking, therefore, in the medium term – say 2000 to 2003:

- If an upward spike is to be avoided in the case where demand remains strong and Iraq does not return fully, more expansion in Saudi Arabia would be necessary to provide spare capacity for emergencies and extreme seasonal demand.
- If a downward spike is to be avoided in the case where there is an economic slowdown affecting demand, and Iraq returns fully to the market, strong quota discipline will be required.

Altogether, the range of uncertainty is wider in the medium than the long term. Signs of further investment in capacity by Saudi Arabia could reduce the upward uncertainty. Signs of the capability of practising stronger quota discipline could reduce the downward uncertainty. Export revenues will be uncertain as usual.

Short term

The prevailing theme in the short term (6–18 months) will be commodity business as normal, but influenced by expectations about the medium term. In late 1997 there were two concerns about the near term, both of which have come to pass: first, demand would fall because of the ramifications of the Asian financial crisis; and, second, Iraq would start to come back to the market, spare capacity would be limited and the risks would be upwards rather than downwards. As of mid-1998, the expectation that sanctions on Iraq could actually be lifted completely has compounded the effects of falling demand in Asia. The risks have therefore been downward, and OPEC producers, with the help of some non-OPEC exporters, have agreed production cuts.

Summary and conclusions

The foregoing discussion of alternative models and available numbers has served to demonstrate how the set of projections which is 'optimistic' on demand for Gulf oil is also the one which is 'pessimistic' about the prospects for sustaining growth in supplies from outside OPEC and the FSU. Clearly this is a reflection not only of the numbers used but also of the internal consistency of the model, and to adjust the projection for demand would inevitably affect that for supply. In fact, as discussed above, it is probable that both facets of the HOPEC projection are open to question on examination of the figures. The view expressed here is that 'non-OPEC ex FSU' supplies will not necessarily peak soon, but other factors will also come into play to affect the place of Gulf producers in the overall energy market. One set of factors explored at length is that likely to influence prices in the long, medium and short term. To conclude, the following summary encapsulates the author's assessment of the overall picture.

About the *long term:*

- Conventional wisdom is growing more confident about economic growth and therefore energy demand.
- There is also more confidence about the ability of non-OPEC suppliers to sustain supply, and of gas to take a larger share of the energy market.
- One cannot have confidence that there will be any fundamental 'escalator' to move oil prices up through the twenties ($20 bbl plus) over the next 15–20 years. There is scope for continuing the historic growth of non-OPEC oil supply, and further penetration of the energy market by gas, as well as further reductions in the cost of converting heavy oils to synthetic fuels or oil products.
- Exporters will have to look to volumes and reduced costs for sustaining increases in revenue.

About the *medium term:*

- With all fluctuations in oil demand and supply being balanced on Gulf exports, prices and export revenues are probably more unpredictable than in the long term.
- Reducing the chance of an upward price spike depends on moves to invest for spare capacity against the contingency that demand continues strong and Iraq does not return fully to the market.
- Reducing the risk of a price collapse depends on demonstrating the ability to exercise quota discipline even if capacity expands faster than demand.

In the *short term*, all the usual uncertainties prevail: anticipation of the medium term will affect market behaviour and temporary shortages or surpluses may have very temporary effects on the market.

3 Development of oil in the Gulf: the UAE in focus

David G. Heard

Some 65 per cent of proven world oil reserves (and 28 per cent of proven natural gas reserves) are concentrated in the Gulf region. The region's share of world oil production, which peaked at 37 per cent in 1973, had declined to about 30 per cent by the second half of the 1990s. Technical factors, as discussed in Chapter 2 by John Mitchell, are only partly responsible for the production-to-reserve ratio. International politics have played a significant part in the history of Gulf oil and continue to do so, not least in accounting for Iraq's near total absence from the market for most of the 1990s.

The intention here is to review the evolution of oil development and production in the region, with special attention to the rise of OPEC and member governments' dealings with the international companies, and focusing specifically on the case of the UAE, in order to arrive at an understanding of the place of the Gulf in the world energy scene as the twentieth century draws to a close. Attention to the case of the UAE, and to Abu Dhabi in particular, demonstrates how regional and international developments frame the context within which government decisions about the energy industry have been made and how Abu Dhabi itself is positioned to respond to the challenges facing all oil producers in the early twenty-first century.

Historical background

The 'Trucial Coast', or the 'Trucial States' were names used by the British to describe the lands of the Arab sheikhdoms on the southern side of the Gulf from the early nineteenth century until the beginning of the

1970s. Historically, European interests in the Gulf were chiefly directed towards the centres of population with water and agriculture, which were found largely on the northern shores. The foreigners, Portuguese, Dutch and then British, were interested in commerce and in establishing safe trade routes from India and the Far East to Europe. This brought them into conflict with the established, and for the most part Gulf national, traders and powers.

The General Treaty of Peace was signed in the year 1820 between the British and the Rulers of the coastal states and Bahrain. The British traders in India as well as Omani and Gulf merchants were expected to benefit from the peace at sea, and importantly peace on the pearl-fishing banks was provided for in the treaty. 'The aim was to minimise possible conflicts of interest between the two parties. Fortunately the interests of the tribes were considered in its formulation.'[1] Further treaties regarding security, political and economic matters followed in later decades until 1971 when, 151 years after the original Treaty of Peace, the British government unilaterally abrogated them.

For most of the period of British predominance in the Gulf region, the British authorities showed little interest in what they regarded as the empty wastes of inland Abu Dhabi, and the barren gravel plains of the northern mountain foreland. The first British explorer to visit the Liwa oases in the south of Abu Dhabi, the old heartland of Abu Dhabi's tribes, was Wilfred Thesiger (Mubarak bin London) who arrived in 1948 by camel from the south. After a short stay he returned across the Empty Quarter to Salalah on the Indian Ocean in the Oman province of Dhofar. He was charmed by the magic of the sands of the Liwa and impassioned by the noble Arab bedouin, but fearful of being caught by a raiding party (*ghazu*).[2]

Initially the British government had no intention of interfering in the internal affairs of the Trucial States but it was drawn in by a desire to keep away potential competitors for trade and influence. The British had growing commercial interests in the Gulf during the early part of the twentieth century, in particular as the British government encouraged British companies to obtain concessions to explore for oil after the First

[1] Frauke Heard-Bey, *From Trucial States to United Arab Emirates: A Society in Transition*, London, Longman, 1982 and 1996, p. 284.
[2] Wilfred Thesiger, *Arabian Sands*, London, Collins, 1983.

World War. The local people of the present UAE expected this greater involvement in local affairs to be accompanied by correspondingly greater economic and administrative assistance. However, the belated British efforts, when they eventually materialized in the 1950s, were judged to be inadequate. This serious omission caused much resentment and great difficulty in later years, a fact which was not well enough understood by the British authorities.

When the break with Britain came in 1971 the exhilaration of some people in what became the UAE was tempered by their awareness of the enormous tasks that lay ahead; others would have preferred to postpone the date. The nation had to be built from a meagre start, and with relatively few able and qualified nationals. Most foreign critics did not rate the chances of survival of this new federal state very highly. Sir Geoffrey Arthur, the last British Political Resident (highest British political authority) in the Gulf, later wrote: 'When the United Arab Emirates was established as a state at the end of 1971 it had few admirers in the West: It was incomplete, it looked loose and ramshackle, and it was born – so said the facile commentators of the day – under the ill star of British patronage.'[3]

A further problem was the external threat to the territorial integrity of the new country posed by disputes over boundaries and frontiers. A specialist in colonial history, J.B. Kelly, stated only a few years earlier that there were very few internationally agreed frontiers in eastern Arabia, and that none of them had been formally demarcated on the ground. He quoted Lord Curzon (Viceroy of India at the beginning of the twentieth century) as saying: 'In Asia there has always been a strong instinctive aversion to the acceptance of fixed boundaries ... more is to be expected from an unsettled than from a settled frontier.'[4] In 1971 the UAE needed stability and settled frontiers both on land and at sea. It is to the credit of the President, H.H. Sheikh Zayed bin Sultan Al Nahayan, and his Foreign Minister Ahmed Khalifa Al Suweidi, that in only a few years a number of boundary agreements were negotiated and signed.

In those difficult early years these leaders ensured that the UAE became internationally accepted, earning a reputation for generous disbursements of aid and gaining a position on the world stage. The UAE

[3] Heard-Bey, *From Trucial States to United Arab Emirates*, p. xxiv.
[4] J.B. Kelly, *Eastern Arabian Frontiers*, London, Faber and Faber, 1964, p. 17.

had to have a very proactive foreign policy, particularly to win acceptance and support from other Arab countries, a few of which were initially quite unsympathetic.

The arrival of the oil industry

From the earliest times the people of the Gulf knew that oil and gas were present in northern Persia and in Iraq. The eternal fires of ancient Iraq, the oil and gas escaping from the surface rocks and catching fire, were known throughout the world. Light crude oil, in Europe distilled from shale oil, had long been used for medical purposes. Where kerosene was available it was used for lighting and fuel: it was cleaner than burning animal fat. Heavy asphalt or pitch was used to make water-tight wooden boats and tanks.

In the early twentieth century, supplies of crude oil in sufficient quantities for transport and industry were discovered in the United States, in the Caspian Sea area and in Burma, but in the uncertain situation before the First World War it was necessary for Europe and the United States to diversify sources.

The first Europeans to risk their money in oil exploration in the countries of the Middle East were men who had become rich financing exploration for minerals such as gold in Australia. They operated from London or Paris paying for work that they never saw and did not understand, yet without their initiative and funds exploration would not have started. Among those remembered in today's oil industry are the pioneer entrepreneurs and tycoons William Knox D'Arcy and Calouste Gulbenkian. Out in the field, the first foreigners and local people to work the primitive machinery lived under very difficult conditions, in constant danger from accidents and disease. Lawlessness in the remote regions where exploration for oil took place added to the instability and discomfort.

Persia was the first country where the sponsors concentrated their efforts because they believed the prospects were good and they were able to obtain concessions to operate. Work had been continuing under great hardship in the western region of Arabistan for some years with only limited success, when in 1908 a message sent from London arrived at the

Table 3.1: Overview of Gulf oil development

	Concession agreement signed	*Discovery*	*Production*	*Depth of reservoir (ft)*
Iran	1901	1908	1912	1180
Iraq	1925	1927	1934	1300
Bahrain	1930	1932	1934	1500
Saudi Arabia	1933	1938	1939	4500
Kuwait	1934	1938	1946	3675
Qatar	1935	1939	1949	5500
Abu Dhabi	1939	1960	1962	8000
Dubai	1937	1966	1969	8500
Oman	1937	1965	1967	–

drilling site with the orders to stop work immediately as there was no more money to support the operations. The chief driller on site, George Reynolds, declared that he did not trust modern telegrams and he would have to wait for a letter containing the order.[5] He continued drilling and found oil before the letter arrived. The oil industry in Persia and the whole of Arabia was born, and fortunes were made thanks to the persistence of this driller who himself returned to obscurity.

After the First World War and the collapse of the Ottoman empire the various international companies – British, Dutch, French and American – joined together, unable to finance such ventures individually, yet eager to negotiate concession agreements with local authorities in the various lands of Arabia. The terms of these initial agreements were considered reasonable for the time – in the years preceding the Second World War – but came into question subsequently. Table 3.1 indicates the order in which concessions were signed and oil came on stream.

Abu Dhabi exploration

The search for oil and gas under the land and sea of the country that was to become the United Arab Emirates started later than in other Gulf countries. In most of the land territory of the Trucial States, except in the

[5] Sir Arnold Wilson, *Letters and Diary of a Young Political Officer 1907–1914*, Oxford University Press, 1942, p. 41ff.

northeast near the mountains of Oman, there were no surface rock structures that attracted the geologist. Nor were there reports of oil or gas leaking up from lower rocks. Prospecting was further discouraged because there were no large natural harbours into which men and heavy equipment could be brought, and there was no easy access across the treacherous coastal Sabkah or over the high impenetrable dunes of the interior.

After long negotiations the first Abu Dhabi concession agreement was signed by the ruler Sheikh Shakhbut bin Sultan in January 1939 with the company which began to develop an interest in the Trucial States, Oman and the entire area. This company, the London-based international, the Iraq Petroleum Company (IPC), whose shareholders were some of the major international companies,[6] had already been producing oil in Iraq since 1934. Witnesses for the ruler were two family advisers, the ruler's uncle Sheikh Khalifa bin Zayed and Ahmed bin Khalaf Al Otaiba. One witness for the company was Ali Tajir, who, in the late 1990s, was living in retirement in Dubai. The agreement[7] included references to minerals as well as oil. There were also clauses that gave some protection to local interests. Similar agreements were signed by the ruler of Dubai and some of the other rulers of the Trucial States.

The economy of the Trucial States had been seriously damaged by the world economic recession of the 1930s. The pearling industry, which had been the mainstay of the economy of the Arab states of the Gulf, had collapsed owing to the introduction of cultured pearls, and the migration of labour to better-paid work in the oil industry in Qatar, Saudi Arabia and Kuwait.

When the war broke out in Europe in September 1939 the oil companies stopped all exploration activities. Abu Dhabi entered a dark era of waiting, enduring considerable hardship. Even after the war, for many years the companies had difficulty finding skilled men as well as obtaining supplies of steel for pipes and other equipment, so they used

[6] The company was initially known as Petroleum Development Trucial Coast. After the discovery of oil in Abu Dhabi, and after concessions were given up in the other Emirates, the company changed its name to Abu Dhabi Petroleum Company. In other areas it was still known as Qatar Petroleum Company or Iraq Petroleum Company. The Abu Dhabi Petroleum Company (ADPC) shareholders to this day are: British Petroleum (23.75 per cent), Shell (23.75 per cent), Exxon and Mobil (23.75 per cent), Total (23.75 per cent) and Partex (5 per cent).

[7] Mana Saeed Al Otaiba, *The Petroleum Concession Agreements of the United Arab Emirates*, London, Croom Helm, 1982.

their scarce resources to develop oilfields further north which they had discovered before the war. The companies sent survey parties back to the Trucial Coast and Oman in 1946, but the prospects were not good. The only significant geological surface features are the few coastal and island salt plugs which have pushed up rocks from deep in the earth. There are no structures visible on the surface, as there are at Dammam in Saudi Arabia, Dukhan in Qatar and Awali in Bahrain, which encouraged drilling in these places.

In Abu Dhabi, where almost the entire territory is covered by sand, the search for oil had to rely on a knowledge of the geology of the region, combined with seismic surveys and test drilling. The first well was sited on the coast between Abu Dhabi and Dubai at Ras Sadr where drilling started in 1950. Work continued for thirteen months with drilling to 13,000 feet, then the deepest well in the Middle East, but after expenditure of £1 million no oil was found and the well was abandoned. Exploration went on for a further ten years before oil was found at Murban. The company also drilled one well in Dubai and one in Sharjah territory.

The company faced innumerable difficult decisions. Working conditions were hard for the foreign employees; and the many nationals had to adapt to a routine which was novel for them. All supplies, including fuel and drinking water, had to be imported. There were no natural harbours large enough for the import of heavy equipment, no airfields and no roads. The coastal Sabkah could not be crossed for months after the infrequent rains and the high sand dunes further inland made travel extremely difficult.

The company faced another very serious problem. Before the possibility of the discovery of oil arose it was not necessary to define land borders, but the arrival of the geologists provoked disputes both among the sheikhdoms and between them and the larger neighbouring countries. The company was not allowed to work in disputed areas, which until 1955 included much of the land of Abu Dhabi. The company played an important role in helping to settle some of these disputes, and thereby to safeguard the territory of Abu Dhabi.

The oil era begins

On 28 October 1960 the company ADPC (Abu Dhabi Petroleum Company; see Figure 3.1) wrote to Sheikh Shakhbut to inform him that oil had been found in commercial quantities at Murban. ADMA (see Figure 3.2), the operator of the Marine concession owned by BP (British Petroleum)

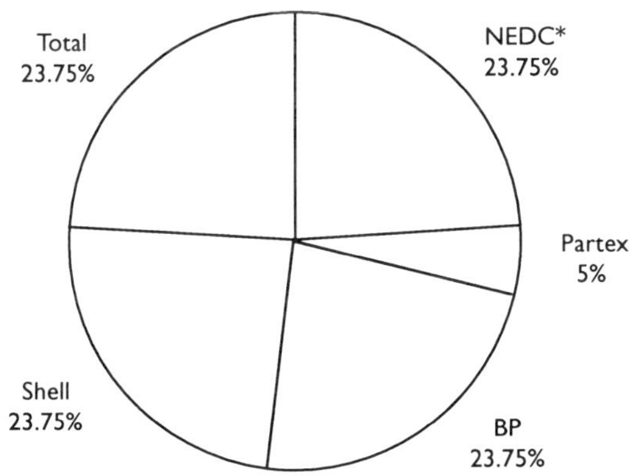

*Near East Development Corp (NEDC) is owned by Mobil (50 per cent) and Exxon (50 per cent).

Figure 3.1: Shareholders in Abu Dhabi Petroleum Company

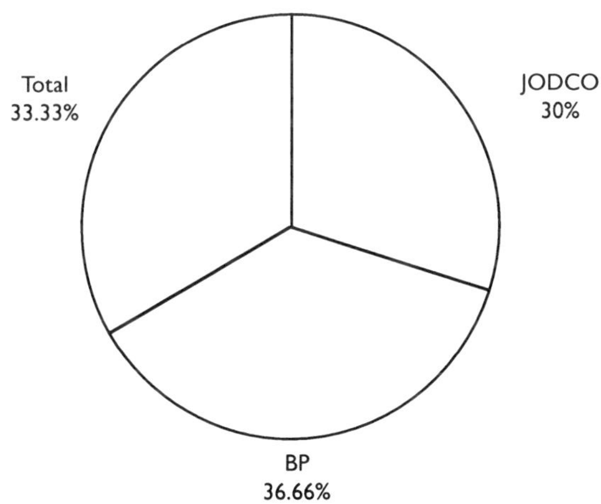

Figure 3.2: Shareholders in Abu Dhabi Marine Areas (ADMA)

in partnership with the French Total, also announced a commercial discovery. ADMA started to export Umm Shaif oil from Das Island at the end of 1962.

By the end of 1963 ADPC was ready to export Murban oil from the Bab field via the newly constructed export terminal at Jebel Dhanna. At first the amounts from both areas were small and because the price at that time was low, very little revenue was generated for government. The companies were selling good-quality Abu Dhabi oil in a very competitive market.

This international oil market was dominated by a few companies, and several of them were represented in ADPC. These companies, based in Europe and America, obviously sought to keep production costs as low as possible in Abu Dhabi so that their shareholders would choose production increases there to meet the growing market demand worldwide in preference to other producers.

The 1965–6 agreements

By the mid-1960s the agreements with ADPC and ADMA had become outdated and unsuitable in the rapidly changing world. In one of his more sympathetic comments about the old agreements, Dr Mana Saeed Al Otaiba says they 'were drawn up unilaterally at a time when the countries of the Middle East were under the influence of the oil companies. It was therefore natural to expect that the agreements in question would be out of step with modern thinking.'[8]

Oil revenues in 1965–6 were very small compared with what was to come, but at the time they were still significant, and the eyes of the world were turning to see how the people of Abu Dhabi and their ruler would react to the new wealth and the promise of unlimited riches to come. The companies were acutely aware that they were open to criticism for maintaining the old agreements, but it took time to explain, negotiate and agree on new terms that would stand up to the scrutiny of critics. After lengthy discussions, Sheikh Shakhbut and the companies finally concluded new agreements in 1965. Among other provisions the basis for

[8] Mana Saeed Al Otaiba, *Petroleum and the Economy of the United Arab Emirates*, London, Croom Helm, 1977, p. 40.

payment to government for oil produced was changed to bring it into line with current OPEC practice. A schedule for relinquishment was agreed, whereby the companies gave up areas to reduce the concession area progressively to the producing fields, allowing the government to explore the relinquished areas or to invite bids from other companies. At the same time an income tax law was passed, applicable only to oil companies and banks.

In April 1965 the Bu Hasa field started to produce for export, thereby more than doubling ADPC exports to over 12m tonnes a year (mta). ADMA was producing 5mta, making a total of 360,000 bpd. Abu Dhabi was becoming a large oil producer.

The accession of Sheikh Zayed

On 6 August 1966 Sheikh Zayed bin Sultan Al Nahayan took over as ruler of Abu Dhabi from his brother Sheikh Shakhbut who, after a short period in exile, returned to live quietly until his death in 1989 in the Abu Dhabi oasis city of Al 'Ain. The new ruler immediately set in motion the processes that led to the rapid transformation of Abu Dhabi to the modern and progressive state that it has become. Oil production rates were still relatively low and revenues were very small by late twentieth-century standards. During the following years considerable pressure was put on the companies to accelerate exploration and development, because Abu Dhabi still had to wait for higher production levels to provide enough money to finance the bold vision of Sheikh Zayed and the expectations of the people.

However, the new ruler had a much more fundamental problem. During the harsh years preceding his accession many people had seen better economic opportunities for themselves and their families in other Gulf countries where oil exports were already generating considerable wealth. Many men or even whole families had gone to Doha in Qatar or Dammam in Saudi Arabia to seek work, education and healthcare, very little of which could be found at that time in Abu Dhabi. Those who remained at home depended on the few and limited government and oil company schools for education and they saw relatively little as yet of what Abu Dhabi might become, while the examples of the rapid development in Doha and Kuwait were well known to them. They were not adequately prepared to build and run a modern state.

People who were educated and who worked abroad had learnt skills that their country urgently required, and they were invited to return to take up positions at home. However, there were not very many of them and the requirements of the new state far exceeded their collective abilities. Therefore the authorities had to seek assistance in almost every field by turning to foreigners. Many people felt very bitter that the previous British authorities and the oil industry had not done more to prepare the people of Abu Dhabi for the time when the inevitable consequences of entering the oil era would have to be faced.

In the event, the efforts of a handful of able people proved crucial in assisting Sheikh Zayed to take the country through a difficult time, which saw not only the emergence of Abu Dhabi as a significant energy producer but also the creation of the state of the United Arab Emirates in December 1971.

OPEC: prices and participation

The Organization of Petroleum Exporting Countries was founded in Baghdad in 1960. The first members were Iraq, Iran, Kuwait, Saudi Arabia and Venezuela.[9] Abu Dhabi joined in 1966, and the UAE in 1974 (see Box 3.1). One of OPEC's first objectives was to fight the decline in oil prices led by Esso's reduction of the price for Arabian light from $1.90 to $1.76 per barrel in 1959. It had been widely accepted for some time both by the producer countries and by the companies that increased producer-government revenues could only be achieved by increases in oil production. The perceived injustice of the dictated decrease provided the incentive for the OPEC countries to cooperate with each other to seek price increases.

In those days the various governments of the OPEC countries were in a weak position. The companies found the oil, they held the expertise, they controlled the markets and they appeared to be all-powerful. During the next few years the companies and OPEC met on many occasions to

[9] See Ian Skeet, *OPEC: Twenty-five years of prices and politics*, Cambridge University Press, 1988.

Box 3.1: Membership of OPEC

Original	*Joined*
Iran, Iraq, Kuwait, Saudi Arabia, Venezuela	September 1960
Qatar	January 1961
Indonesia, Libya	June 1962
Abu Dhabi (January 1974 UAE)	December 1966
Algeria	July 1969
Nigeria	July 1971
Ecuador	November 1973
Gabon	July 1975

discuss and implement small price increases, but the atmosphere was not good, being influenced, at least in part, by the international politics of the time. By the early 1970s the pressure was on the companies to give more, and they gave very reluctantly in small amounts. Agreements were reached and quickly superseded by events; small increases in prices were agreed and they held for a short time. Abu Dhabi enjoyed a premium for its low-sulphur oil, others for the attractive gravity of the oil or relative proximity to the market.

By 1973 OPEC members had gained more confidence and experience in negotiating with the companies. They were preparing for yet another meeting on 8 October 1973, this time in Vienna, when, on 6 October, Egyptian forces crossed the Suez Canal and the third Arab–Israeli war was under way. At the meeting in Vienna, the companies made their offer, which in the eyes of the OPEC negotiators was totally inadequate. The counter-offer from OPEC went far beyond the competence of those representing the companies and they packed their bags and returned home. On 16 October the OPEC team assembled in Kuwait and unilaterally announced an increase of $2 per barrel. They had seized the initiative and from then on pricing was no longer in the hands of the companies.

While the oil ministers and energy companies were reappraising their positions, Arab politicians' primary concern was about Arab loss of life in the Arab–Israeli war. President of the UAE Sheikh Zayed was quoted as saying that 'Arab oil is not more valuable than Arab blood', and he was one of the first to advocate and implement the boycott of oil exports to

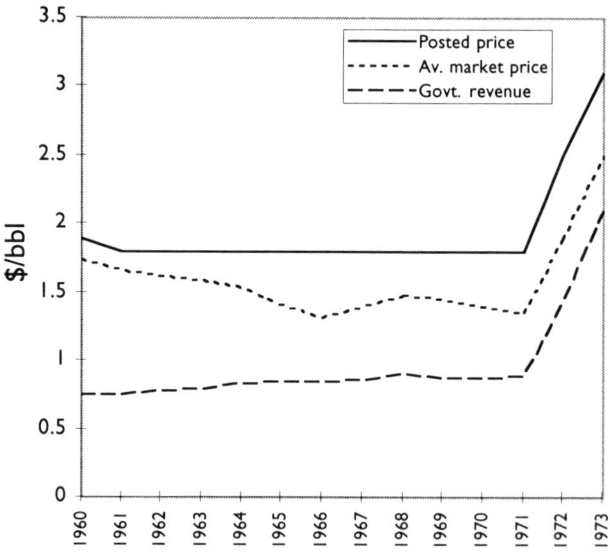

Figure 3.3: Oil prices, 1960–73

countries seen to be supporting Israel. At the November OPEC meeting the Arab members agreed on a total embargo on shipment to certain countries, in particular the United States and the Netherlands, as well as production cuts increasing at 5 per cent each month. The result was that oil prices soared on the spot market and non-Arab OPEC members saw themselves selling oil at an unprecedented $20 a barrel (see Figure 3.3).

The OPEC countries were not only seeking price increases. By the end of the 1960s there was an increasing demand among members, in particular those with mature oil industries and adequate governmental infrastructure, for a say in how their most important natural resource was disposed of. Governments hoped to take control of the industry in their lands for themselves, but they felt powerless in front of the companies, while their people were deeply resentful of the foreign presence which they saw as domination. It was feared that a producer country that nationalized the interests of some of the oil companies would be excluded from the world market by the all-powerful industry. In fact many countries believed that they needed the cooperation of the companies; even Iraq hesitated to nationalize the whole of IPC in 1972, but started by partial nationalization of IPC, leaving its BPC (Basra Petroleum Company) arm intact.

The opportunity for OPEC governments to make a fundamental change in their relationships with the oil companies resulted from a number of events which placed different pressures on the companies and producer countries and thus led to an increasing diversity of objectives among them. The scene was set thus:

- The closure of the Suez Canal in 1967 put Gulf crude oil at a price disadvantage when compared with North African or West African crude in the Western market. The introduction of giant supertankers, up to 1 million tonnes, helped to restore this balance several years later.
- The Ba'ath party came to power in Iraq in July 1968, excluding the possibility of an improvement in this country's bad relations with the London-based Iraq Petroleum Company.
- In September 1969 there was a revolution in Libya which brought the military strong man Muammar Qaddafi to power.
- The Nigerian civil war ended in 1969 and the country needed revenue to recover and rebuild.
- The announcement of the British withdrawal from the Gulf caused some instability, followed by Iran's occupation of the three islands over which it remains in dispute with the UAE.
- During the last three years of the 1960s world growth in demand for oil was exceptionally high.
- The Saudi Arabian Oil Minister Sheikh Zaki Yamani and the Kuwaiti Oil Minister Atiqi discussed 'participation versus nationalization' in public.
- The American oil company, Occidental, was a very large oil producer in Libya, but had relatively little production elsewhere, which made it an easy target for the hawkish Libyans.
- In January 1970 Algeria became the first OPEC country unilaterally to declare a price for Algerian oil after long negotiations with the French had failed.
- In August 1971 uncertainty was created when the United States suspended the connection of the dollar with gold priced at $32 per ounce.
- In June 1972 IPC in northern Iraq was nationalized, which caused more turmoil.

Throughout 1972 discussions and negotiations on participation continued. Participation, whereby governments took a shareholding in a company's concession, rights, obligations and operations in their countries, was seen as a workable compromise avoiding the uncertainties associated with nationalization. The outcome was a mixture of nationalization and participation, and though not as dramatic as the price upheavals that were to come, it had much more far-reaching effects. The Participation Agreement, effective from 1 January 1973,[10] allowed the producing countries to take a 25 per cent share in the operations, which would rise each year by 5 per cent until it reached 51 per cent. Compensation was to be paid to the companies, and the companies were committed to assist with the marketing of the oil. Since most of the OPEC countries did not accept this 1973 agreement and were adopting various positions particular to their own interests, a period of uncertainty ensued culminating in individual arrangements.

Abu Dhabi negotiated a further agreement effective from 1 January 1974, in which the government acquired a 60 per cent share, and the companies retained 40 per cent.[11] (Saudi Arabia, Iran, Qatar and Kuwait chose nationalization.) After the signing of the general participation agreements, the companies ADPC and ADMA continued to operate the concessions on behalf of the Abu Dhabi National Oil Company (ADNOC) and the foreign shareholders. Agreements specific to each of the two operations were negotiated over the next few years, culminating in the establishment by decrees in 1978 of the two operating companies, Abu Dhabi Marine Areas Operating Company (ADMA-OPCO)[12] and Abu Dhabi Company for Onshore Oil Operations (ADCO) (see Figures 3.4 and 3.5).[13] By 1 January 1979 the processes were complete and the transition to the new companies passed without any major difficulties.

[10] The Participation Agreement was signed on 20 December 1972 in Riyadh by Saudi Arabia and Abu Dhabi with the major oil companies operating in the two countries: Exxon, Atlantic Richfield, British Petroleum, Compagnie Française des Pétroles, Gulf Oil, Mobil, Partex, Shell, Standard Oil of California and Texaco. Qatar signed in April 1973.

[11] The 1974 Participation Agreement was signed on 2 September 1974 in Abu Dhabi by the Government of Abu Dhabi with the Abu Dhabi Petroleum Company and British Petroleum, Compagnie Française des Pétroles (Total), Exxon, Mobil, Partex and Shell. The equivalent ADMA agreement was signed on 13 September in Abu Dhabi by the Government of Abu Dhabi with Abu Dhabi Marine Areas and British Petroleum, Compagnie Française des Pétroles and Japan Oil Development Company Limited.

[12] Law No. 6 of 1977.

[13] Law No. 14 of 1978.

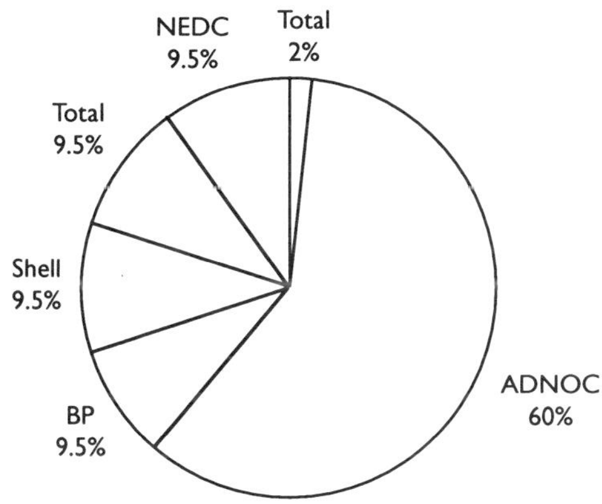

Figure 3.4: Shareholders of ADCO

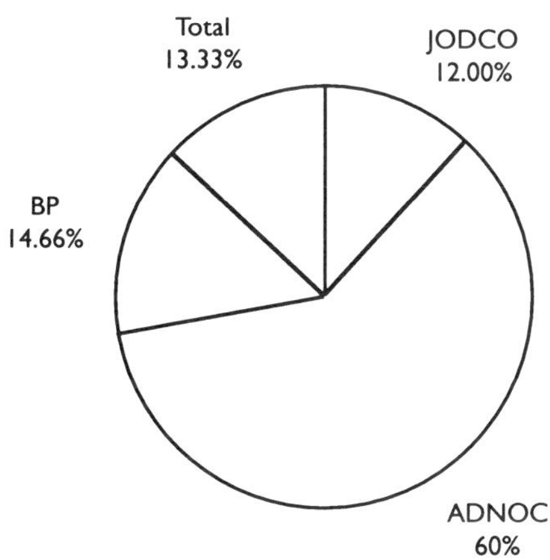

Figure 3.5: Shareholders of ADMA as of 1978

Oil has become political

Although the oil companies were obliged, as time went by, to provide governments with more information about their operations, the total amount given was very little in the early years. Not only in Abu Dhabi but throughout the region, governments were ill informed about even the most basic facts that they needed to have in order to formulate their oil and financial policies. The governments could read data for their reserves of oil and gas in the oil press, but they were not given the information by the companies to derive them for themselves. Even in the early 1970s when large amounts of raw data were passing from the companies to the governments, there was a total embargo on any information that could be called 'interpretative'. The companies set oil production levels, determined exploration programmes and decided whether and when to develop new finds. The companies were very secretive and were perceived to be arrogant; this combination bred mistrust. The companies presumably had very good commercial reasons for being secretive, but politically it was a short-sighted approach.

In the Six-Day Arab–Israeli war of 1967 OPEC stood aside and did not take any action. The partial ban on exports to 'all countries helping Israel' was decided by the Arab League and implemented by its oil-exporting members. A weakness in OPEC that was to plague it later was already obvious: there was no way of compelling member states to observe collective agreements which they did not support. This same weakness also affected the so-called Seven Sisters, the large oil companies. In complete contrast, in October 1973 the Arab members of OPEC were leading the action and, as described above, foremost among them was the UAE led by Sheikh Zayed.

OPEC and Abu Dhabi

Abu Dhabi joined OPEC at the end of 1966, initially represented by the very competent Iraqi Dr Nadim Pachachi (1966–71), who was also OPEC Secretary General during 1971–2. Soon Dr Mana Saeed Al Otaiba took over oil affairs in Abu Dhabi, and he remained UAE Minister of Petroleum and Mineral Resources until 1990. Under the direction of Sheikh Zayed, he led the negotiations which resulted in Abu Dhabi increasing its interest in the ADPC and ADMA concessions to a 60 per cent equity share in the 'rights, operations, assets, obligations and facilities of the main oil producing activities in Abu Dhabi'. The historic General Agreement on Participation was signed on 2 September 1974 and took effect from 1 January 1975.

The issue of participation was thus cleared from the table, but during the coming years price and allowable production levels presented major problems, and Dr Al Otaiba was hard put to defend the best interests of his country. On numerous occasions he called for discussions recognizing that these problems presented a threat to producers which was every bit as serious as the threats faced when OPEC was first established. Abu Dhabi (and then the UAE) played a very active role in OPEC, in general supporting Saudi Arabia and other Arab countries when individual members took different positions.

Abu Dhabi government and the oil companies

In Abu Dhabi, notwithstanding their differences, the relationship between government and the companies has benefited from a spirit of cooperation through consultation. Sheikh Zayed formally started to delegate authority when he established a government Department of Oil Affairs in 1969, but he continued to take a personal interest in the day-to-day affairs of the industry until well into the 1970s. He appointed the young Baghdad-educated national, Mana Saeed Al Otaiba, to the department, initially as deputy head, and he reported directly to Sheikh Zayed during his whole period in office.

The Department of Oil Affairs, later called the Department of Petroleum and Industry, was the government vehicle for safeguarding its interest in matters concerning the oil concessions, policing the concessionary companies and ensuring that they met their obligations under the various agreements. The department dealt with Abu Dhabi's oil-related matters, such as the award of new concessions to explore for oil, as opposed to the UAE Ministry of Petroleum which dealt with the UAE's international oil affairs, including OPEC. Dr Al Otaiba served as chairman of the department until it was dissolved in 1988 and the Supreme Petroleum Council was established.[14]

Abu Dhabi National Oil Company

ADNOC was founded in November 1971: 'To engage in all stages of activity in the Petroleum Industry in Abu Dhabi and outside.'[15] The company would also have a number of affiliates and subsidiaries. The ADNOC

[14] Law No. 1 of 1988.
[15] Law No. 7 of 27 November 1971.

board was recognized as the highest authority for petroleum affairs in Abu Dhabi until the Supreme Petroleum Council was set up and supplanted it in June 1988. ADNOC is the majority shareholder in the three main oil-producing companies, the two gas-processing companies and six oil-related companies, and is the owner of three oil-related companies.[16] (For further information on company holdings see Appendix at the end of this chapter.)

After participation to 1980

As the OPEC governments and the oil companies were assimilating their new and sometimes uncomfortable relationships, and nationalization or participation ceased to be such a hot issue, the price of oil entered a period of nervousness followed by instability. For the next few years OPEC ministers discussed and argued over price and price differentials, with Iran and Iraq in particular trying to accelerate the gradual price increases, while Saudi Arabia urged restraint. The differentials were amounts added to or subtracted from the basic price to reflect differences such as crude oil quality. There were those in OPEC who had their eyes on a long-term strategy, but discussion of this was stifled by bickering over immediate short-term and pressing issues.

At the end of 1974, the US-inspired International Energy Agency (IEA) was set up, in the face of opposition from France and suspicions among OPEC oil ministers, but with the support of the main oil-consuming countries. As it transpired, IEA did not become an effective focus for dialogue between consumers and OPEC. Meanwhile, the United States resorted to more direct and personal pressure on some governments to try to persuade them to accept price restraint. For their

[16] ADNOC holds 60 per cent of ADCO (Abu Dhabi Company for Onshore Oil Operations); 60 per cent of ADMA (Abu Dhabi Marine Areas Operating Company); 88 per cent of ZADCO (Zakum Development Company); 70 per cent of the National Petroleum Development Company; 75 per cent of the Abu Dhabi Drilling Chemicals and Products Company; 100 per cent of the National Drilling Company; 60 per cent of the Abu Dhabi Petroleum Ports Operating Company; 60 per cent of the National Marine Services Company; 68 per cent of GASCO (Abu Dhabi Gas Industries Company); 70 per cent of ADGAS (Abu Dhabi Gas Liquefaction Limited); 66.66 per cent of Ruweis Fertilizer Industries; 100 per cent of the Abu Dhabi National Tanker Company; 51 per cent of the Liquefied Gas Shipping Company; and 100 per cent of the Abu Dhabi National Oil Company for Distribution.

part, the major oil companies, deprived of their main sources of crude oil through nationalization, sought alternative reserves over which they could exercise greater control and which would help them to maintain a measure of independence. The increase in oil prices in the 1970s meant that many new opportunities around the world became economic.

Against this backdrop, for a brief period OPEC seemed to be moving towards greater internal cohesion, prices appeared to be steadying somewhat and the post-participation relationships were settling into place. However, new and very serious events were taking place in the Middle East that contributed to instability and to the perception that the area was not a reliable source of crude oil.

- King Faisal of Saudi Arabia was murdered on 28 March 1975.
- The international terrorist Carlos kidnapped OPEC oil ministers on 21 December 1975 while they were in session in Vienna, releasing them two days later in Algiers.
- In 1975 Iran and Iraq signed an agreement, which lasted only four years, to settle their border dispute.
- In September 1978 the Camp David accords were signed by Egypt and Israel, isolating Egypt from the Arab world and causing disputes among the Arab states.
- In the summer of 1978 there were strikes in the Iranian oil installations: observers correctly interpreted these as the start of a new upheaval.
- In January 1979 the Shah of Iran was forced out. Iranian exports became unpredictable, varying from little to normal. The uncertainties took the price of oil on the spot market to $40 per barrel and delivered a shock to the world economy. Saudi Arabia and the UAE urged moderation as all could see that the conservation measures introduced in the main purchasing countries were reducing consumption.
- On 4 November 1979 the US Embassy in Tehran was seized and hostages were taken.
- On 20 November 1979 the Muslim holy shrine, the Grand Mosque in Mecca, was seized by a group of terrorists.
- On 27 December 1979 the Russians invaded Afghanistan.
- In September 1980 the Iran–Iraq war started, seriously reducing the amount of oil on the market and further alarming the consumers.

Turning point for OPEC

During the latter half of the 1970s OPEC apparently benefited from the growing instability, witnessing a rise in the price of oil from $12 to $32 bbl over a period of five years. In November 1977 Abu Dhabi decreased oil production by 12 per cent to 'preserve some of the major reservoirs, after studies made by ADNOC'.[17] Further cuts followed in 1979 and 1980 for the same reason. A turning point for OPEC was reached in 1981, although cumulative consumer market cutbacks were concealed by the loss of Iranian and Iraqi production, and by year end a marker price of $34 bbl was agreed. In fact, the whole nature of the world market had changed. The value of exports from OPEC countries, including those of Abu Dhabi, had peaked in 1980. (For UAE production and OPEC price trends see Figures 3.6 and 3.7.) Soon the loss of Iranian and Iraqi production was not enough to sustain the high price and before long urgent needs within some OPEC countries, such as military contingencies and defence costs, meant that very substantial discounts were being offered to move oil in the declining market.

When OPEC met in March 1983 in London, oil production in Saudi Arabia and Abu Dhabi had fallen to half the previous peak levels. The oil ministers sought stability in the market and recognized that this could be achieved only by price reductions.[18] But in spite of moderate price cuts, by 1985 Saudi Arabian production had fallen from over 10m bpd to less than 4m bpd.

Pressure on prices was coming from a new direction. The companies which had lost direct access to Middle East oil had found and developed new reserves worldwide in countries which were not OPEC members. OPEC's share of the 'free world' crude production had declined from 65 per cent in 1976 to 40 per cent in 1985. While Middle East production was being severely cut, production doubled in western Europe, rose significantly in Latin America, rose 20 per cent in the FSU and was maintained in North America. The result was a gradual weakening of the oil price, which fell from above $30 to below $20 bbl.

[17] ADNOC *Annual Review 1978*, p. 14.

[18] 'OPEC's price raising activity in 1979–80 gave a powerful fillip to the development of high cost production in non-OPEC areas ... since these high cost producers must be allowed by their governments an adequate return on their investment, OPEC has automatically condemned itself to a shrinking share of world production.' *Petroleum Economist,* December 1983, pp. 450–51.

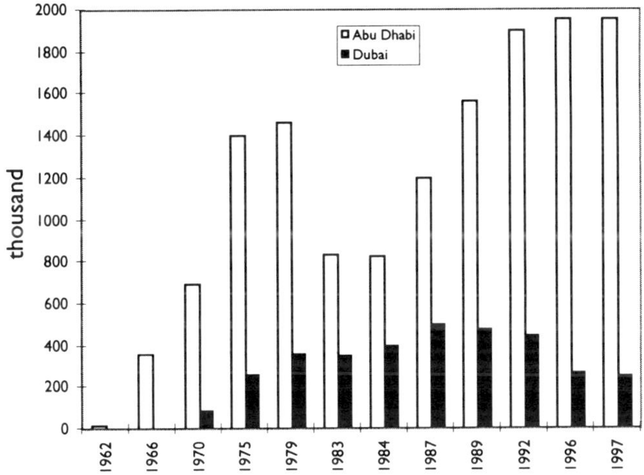

Figure 3.6: UAE oil production levels, 1962–97*

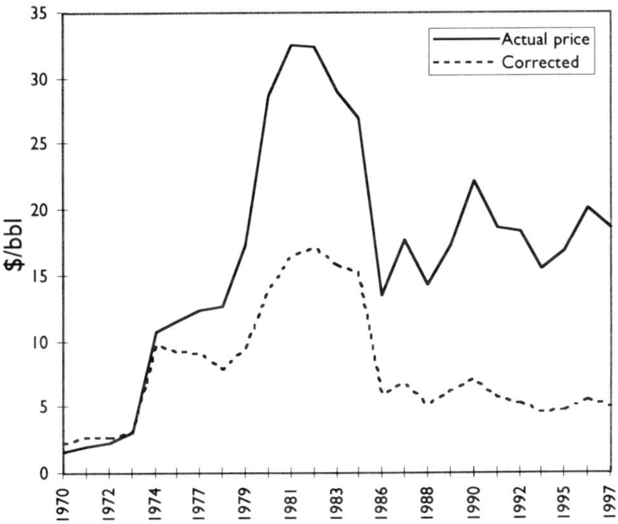

Figure 3.7: OPEC crude oil prices, 1970–97

At the end of 1985 Saudi Arabia, which had borne the brunt of the production cuts, decided that it had to try to raise revenue by increasing production. The price of oil fell by half, to less than $10 a barrel in August 1986. Abu Dhabi production had already crept up from the low 1984 level.

* Data for this and subsequent figures derived from various sources, including *Middle East Economic Survey*.

OPEC production quotas

Meeting in December 1986, OPEC members declared the need to return to stable prices and they agreed to limit production. Figure 3.8 provides a demonstration of the considerable fluctuations over time in the value to OPEC members of their oil exports. Figure 3.9 demonstrates the same for the Arab oil exporting countries (OAPEC). The UAE was given a quota of 902,000 bpd for the first quarter of 1987, rising to 1,045,000 bpd in the last quarter. At its September 1987 meeting the OPEC committee responsible reported widespread 'non-compliance' with the agreed quotas.[19] There was also a veil of secrecy imposed by individual countries on reporting production levels.

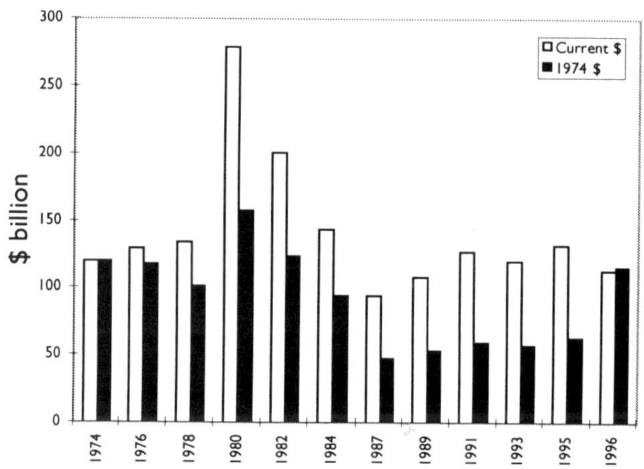

Figure 3.8: Value of OPEC oil exports

For the next three years OPEC was preoccupied with production quotas. A market monitoring committee was established, then there were attempts to introduce new criteria for deciding appropriate production quotas. One argument which gained currency was that countries with higher oil reserves in the ground should be permitted higher quotas. As a consequence, several states re-examined and adjusted their reserve figures.[20]

[19] 'Total OPEC production ... was in excess of the agreed ceiling, although the excess is well below what has been reported by the specialised media', *MEES,* 14 September 1987.
[20] 'Dr Mana Saeed Al Otaiba announced 200 billion barrels of oil in place, a 50% recovery factor, giving 100 billion barrels recoverable reserves.' These figures are roughly three times higher than previous estimates. *MEES,* 2 November 1987.

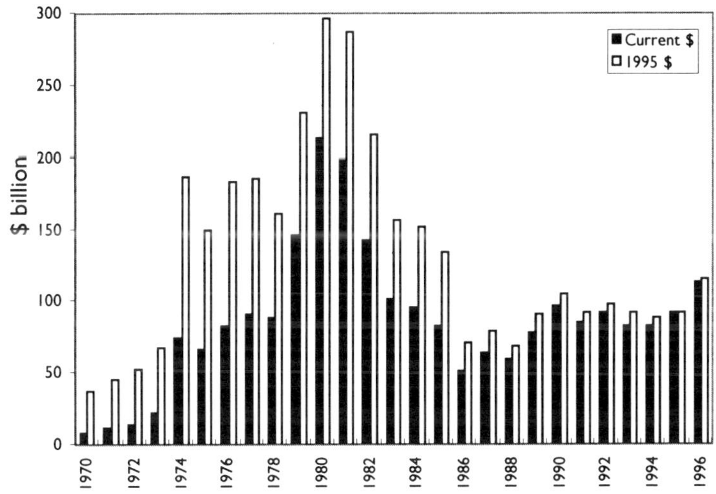

Figure 3.9: Value of OAPEC oil exports

The UAE failed for several years to win a significantly larger OPEC production quota, but in the prevailing climate of secrecy it was able to let exports slide upwards. After strenuous efforts Dr Mana Al Otaiba managed in July 1990 to achieve an increase in the OPEC quota for the UAE from 1,095,000 bpd to 1,500,000 bpd.[21] When Iraq invaded Kuwait on 2 August 1990 higher UAE exports as well as increases from Saudi Arabia and other countries helped to stabilize the market during a difficult few months. By the end of the year Abu Dhabi was reported to be producing over 2m bpd to meet demand arising from the loss of oil from Iraq and Kuwait.

During the mid-1990s the UAE held to the quota of 2,161,000 bpd allocated in agreement with OPEC in September 1993. (The production of condensate and gas is not directly limited by the quota.) This helped to achieve the objective of price stability into 1997.

The contemporary position of the UAE

The UAE is reported to have under its land and waters some 9 per cent of the world's crude oil reserves and 4 per cent of its natural gas reserves (see Figures 3.10 and 3.11).

The UAE accounts for 3.7 per cent of the crude oil and 0.7 per cent of

[21] *MEES*, 22 April 1991.

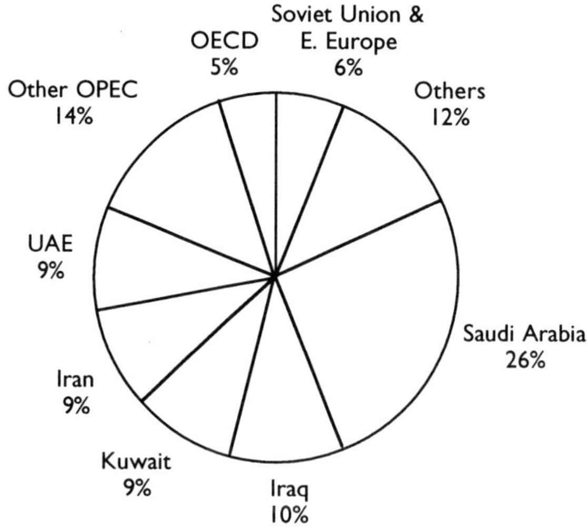

Figure 3.10: World proven oil reserves

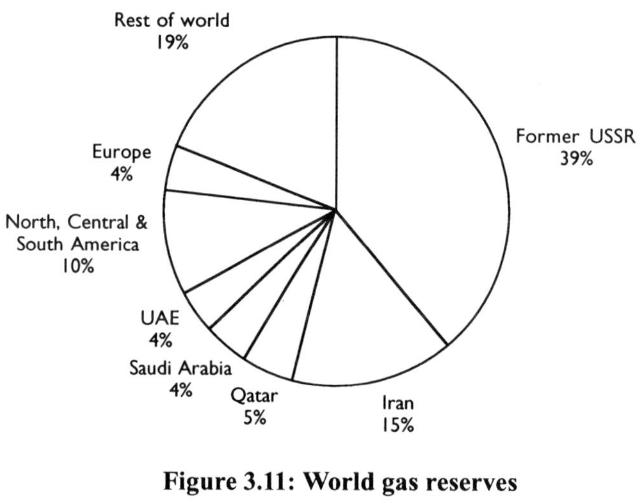

Figure 3.11: World gas reserves

the natural gas production of the world today (see Figures 3.12 and 3.13).

Within OPEC, the UAE accounts for 8.5 per cent of total OPEC production, and its share of the total production from the Arabian Gulf Cooperation Council (GCC) countries is around 16 per cent (see Figures 3.14 and 3.15).

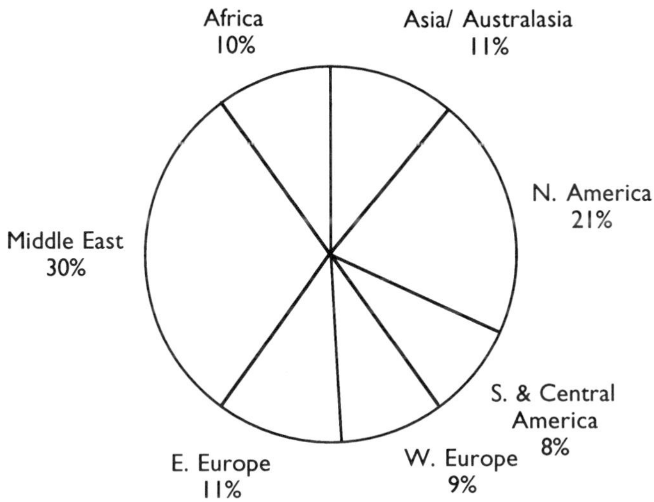

Figure 3.12: World oil production

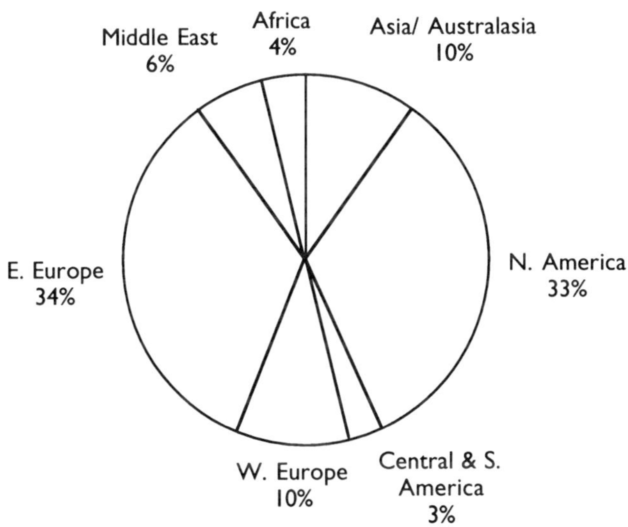

Figure 3.13: World gas production

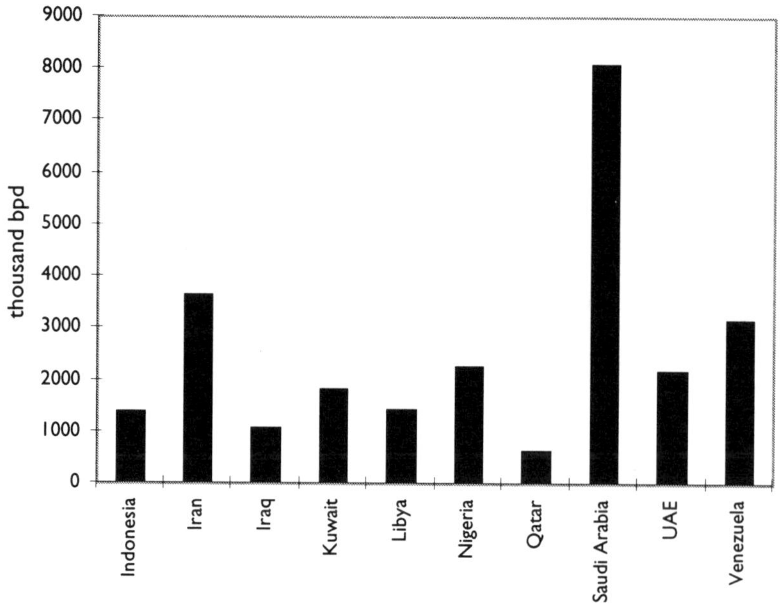

Figure 3.14: OPEC crude oil production 1997

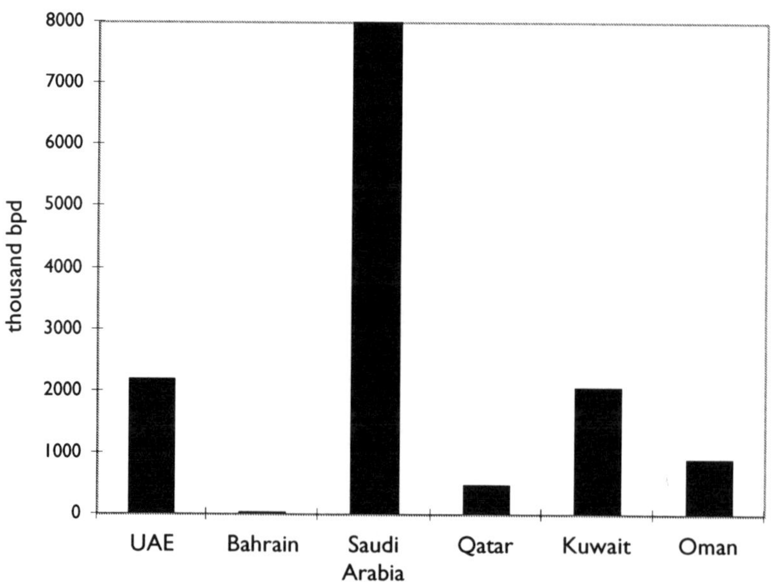

Figure 3.15: GCC production 1997

As discussed above, the UAE was a comparative late-comer to Gulf oil production, commencing the export of crude in 1962. The amounts were at first quite small, but exports increased as new fields were discovered and brought on stream. UAE commercial gas exports commenced in 1977 and by 1982 almost all gas produced with the oil (associated gas) was being used or exported. The first refinery at Umm An Nar commenced operations in 1976, and the larger Ruweis refinery in 1981. As of 1998, 10 per cent of the UAE oil production passes through these two Abu Dhabi refineries.

The relative position of the UAE among other GCC producers in 1997 is illustrated in Figure 3.15. In future years the increasing effort required to maintain oil production will translate into rising production costs. There will also be very large investments in natural gas and petrochemicals projects.

Within the UAE, the oil and gas reserves of the smaller emirates, Sharjah and Ras al Khaimah, are dwarfed by the quantities in Dubai, which in turn are very small when compared to the amounts in Abu Dhabi. Dubai commenced exporting oil from the offshore Fateh field in 1969.[22] The production rate rose to over 400,000 bpd (less than one-quarter the Abu Dhabi rate) and has been declining for several years. The original oil reserves of Dubai are reported to have been about 4 per cent of the reported Abu Dhabi reserves. There are several Dubai oilfields, all in the sea and near the median line with Iran. So far efforts to halt the decline in production have been very costly. Dubai also produces some gas from the small onshore Margham field. Sharjah exports oil from the Mubarek field beside the island of Abu Musa. Production, which reached about 8,000 bpd, has declined. The onshore gas fields continue to produce for local consumption in Sharjah and Dubai. Ras al Khaimah had small offshore production of oil for only a few years.

There have been no major oil or gas discoveries in the territory of the UAE since Zarara was discovered in 1971. This field is the northwards extension of the Shuaiba field which is currently being developed in Saudi Arabia. Work has continued throughout the land exploring and appraising discoveries. It is highly likely that many more relatively small

[22] Shareholders in Dubai Petroleum Company are: Conoco 32.5 per cent, Total 27.5 per cent, Repsol 25 per cent, RWE-DEA AG 10 per cent, Wintershall 5 per cent.

discoveries will be made in the future, but none of them will be as large as the giants found many years ago. The largest and most significant uncertainty for the future is the size of the oil reserves. The amount of oil in the ground can be determined fairly accurately, but how much will be produced depends on wise management of the reservoirs and the employment of the most advanced technology available.[23]

The value of oil exported may rise with time, but the revenue accruing to the government in the longer term will be diminished by the increasing cost and complexity of oil operations. Meanwhile, in recent years the economy of the UAE has seen industry, trading and investments grow in importance to diversify away from total reliance on oil and gas exports.

Conclusion

The story of the past thirty years shows that political instability in the Gulf is damaging to the oil producers there. Every time the sword is unsheathed the world is reminded of this instability and increases its efforts to secure alternative sources of energy, oil and gas. This helps to account for the decline in the Gulf region's share of world production, even though it possesses the highest concentration of reserves.

Within this context, however, the UAE does enjoy certain advantages, in part because the country was a comparative latecomer to energy development and was not well prepared for the challenges attending statehood in 1971. As it transpired, the Emirates progressed along a path which combined the establishment of state structures with a policy for the energy sector which proved both pragmatic and far-sighted. The consistent and firm leadership of Sheikh Zayed, for more than thirty years, has provided the continuity and stability essential for building and developing the political, social and economic structures. The policy of

[23] The *Petroleum Handbook* published by Shell discusses recovery factors. With light oil, employing primary processes between 10 and 40 per cent of original oil in place may be recovered, and a further 10–25 per cent by secondary processes. For medium oil the figures are 5 and 30 per cent and 5–15 per cent respectively. The Norwegian Petroleum Directorate is reported to say '... shows a rise in the average recovery factor to 41 per cent from 34 per cent... The NPD has set a target ... of 50 per cent recovery for oil based on expectations of further technological advances.' *Petroleum Economist*, August 1997, p. 8.

retaining the foreign shareholders in the energy sector, albeit as minority partners, has contributed to stability in the industry. The partners represent the largest and most experienced oil companies in the world, yet their continued role in the development of Abu Dhabi's oil, gas and related industries reflects their acknowledgment of the principal objective of maximizing the benefit to the host country of its natural resources.

The views expressed here are those of the author and do not necessarily reflect the views of either the government of the UAE or the Abu Dhabi Petroleum Company.

Appendix: The UAE – Energy-related structures and companies

The UAE Minister of Petroleum and Mineral Resources acts for Abu Dhabi, Dubai and all the Emirates in appropriate matters such as representation at OPEC and OAPEC.

The Supreme Petroleum Council (SPC) under the Chairman Sheikh Khalifa bin Zayed Al Nahayan, Abu Dhabi Crown Prince and Deputy Commander in Chief of the UAE Armed Forces, is the highest petroleum authority in Abu Dhabi. The SPC implements government policies and protects the interests of the government. It advises monthly in arrears the oil price to ADNOC and all lifters of crude oil.[1]

The International Petroleum Investment Company (IPIC) is responsible for Abu Dhabi's overseas investments in the oil industry. Its main investments are in the Spanish Repsol, the Austrian OMV, the Danish Borealis, the Pakistani Multan refinery and the Egyptian Suez Mediterranean pipeline.

ADNOC, originally founded in 1971, is both directly and indirectly involved in all aspects of the industry in Abu Dhabi. Where it shares ownership of organizations it is the majority and dominant shareholder pursuing the best interests of the state of Abu Dhabi. A breakdown of all the companies operating in Abu Dhabi appears in Table A3.1.

ADNOC is engaged in all aspects of oil and gas exploration and production, with the foreign partners and on its own account. ADNOC has a number of oil and gas discoveries to its credit. It sells crude oil and gas on the world market and engages in all aspects of the oil industry, either directly or through its subsidiaries, and is moving into petrochemicals.[2] In April 1976 the first refinery in Abu Dhabi, wholly owned by ADNOC, was commissioned. By August 1983 refining

[1] Different prices are advised for different qualities and specific gravities of crude oil.

[2] In May 1996 the Supreme Petroleum Council approved the award to Borealis (40 per cent) in partnership with ADNOC (60 per cent) of the project to set up a polyethylene plant in Abu Dhabi at Ruweis.

Table A3.1: Shareholders in Abu Dhabi main oil and gas operating companies

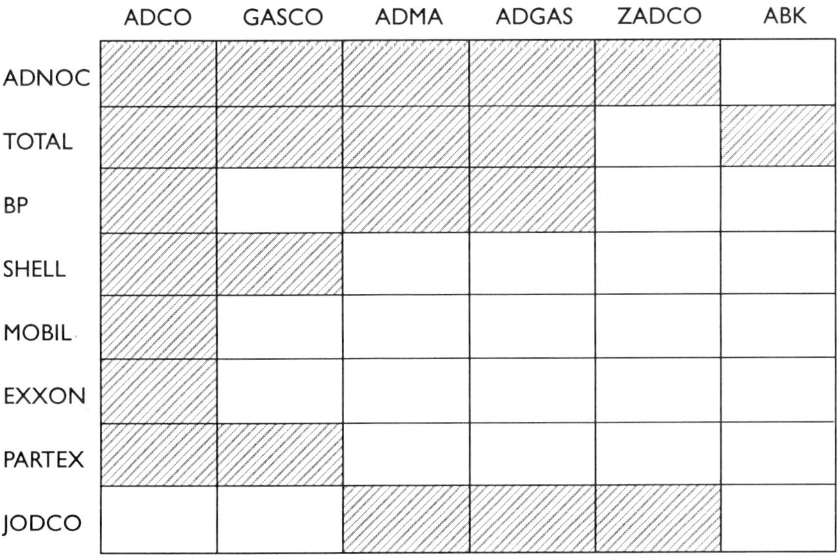

capacity had risen to 180,000 bpd. The government of Abu Dhabi took possession of all gas including associated gas in 1976.[3]

In 1972 Dr Mana Saeed Al Otaiba signed the agreement to establish ADGAS[4] (see Figures A3.1 and A3.2) to handle offshore associated gas and to sell the product to the Tokyo Electric Power Company. Exports commenced in 1977 when the first LNG[5] plant in the Gulf came on stream.

In June 1977 Sheikh Tahnoon bin Mohamed Al Nahayan, chairman of ADNOC, signed an agreement with the French company Total for the development of the Upper Zakum oilfield. This field had been produced at a low rate by ADMA, and ADNOC wished to develop the vast reserves further.[6] In July 1978 Sheikh Tahnoon bin Mohamed signed the agree-

[3] Law No. 4 of 1976.

[4] Shareholders in ADGAS were ADNOC (51 per cent), Mitsui (24.5 per cent), BP (16.33 per cent) and Total (8.17 per cent), and from June 1997 they owned 70 per cent, 15 per cent, 10 per cent and 5 per cent respectively.

[5] LNG is cooled to below minus 161.5 degrees C.

[6] ADNOC exercised its rights under the terms of the Participation Agreement to develop the Upper Zakum field, a complex and very large structure. BP and Total declined. The remaining participating parties are ADNOC (88 per cent) and JODCO (12 per cent).

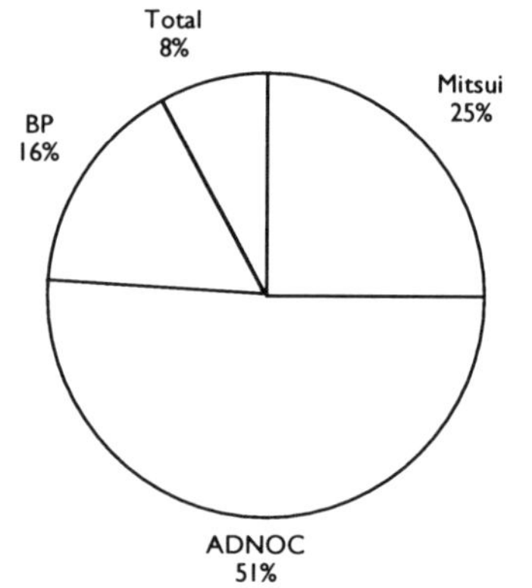

Figure A3.1: Shareholders of ADGAS, 1972

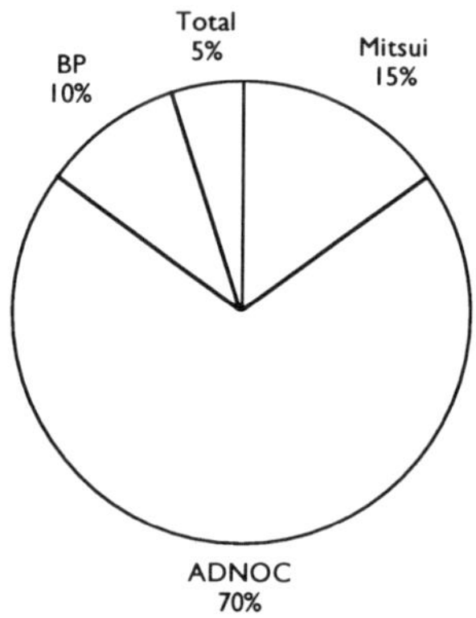

Figure A3.2: Shareholders of ADGAS, 1997

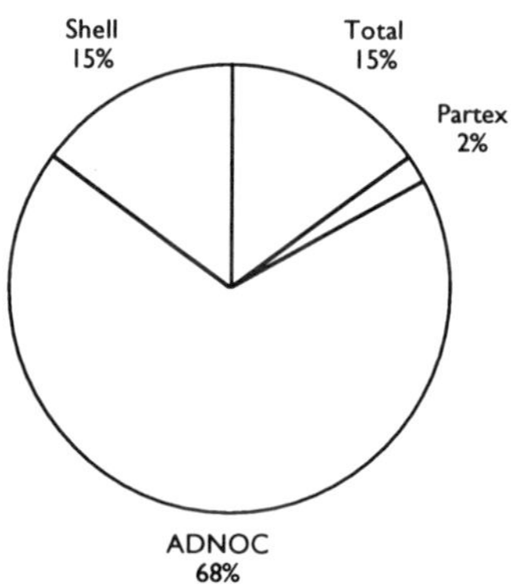

Figure A3.3: Shareholders of GASCO

ment for the establishment of GASCO[7] (see Figure A3.3) to handle onshore associated gas. Exports began in 1981.

In April 1987 Sohail Faris Al Mazrui became acting general manager of ADNOC. He was confirmed in this position in June 1988, when the Supreme Petroleum Council was established as the ultimate authority in Abu Dhabi oil and gas affairs.[8] In November 1994 Yousef Omeir bin Yousef became general manager of ADNOC.

The operating companies in Abu Dhabi

- Abu Dhabi Company for Onshore Oil Operations (ADCO).
- Abu Dhabi Marine Areas Operating Company (ADMA-OPCO).
- Zakum Development Company (ZADCO).

The government has exercised control through ADNOC over the oil operations since the Participation Agreements of 1974, whereby it acquired a majority shareholding, and the implementing agreements of 1977–8 setting up the operating companies.

The agreements established the Joint Management Committee (JMC) as the highest authority for each of the oil operating companies. Members of the JMC are the government with ADNOC, the concession holder and the foreign shareholders. Below the JMC is the board of directors composed of representatives from all shareholders (ADNOC and the foreign shareholders). The Finance and Technical Committees, supported by sub-committees and technical workshops, report to the board. These committees are all composed of technical experts from all shareholders and they meet from time to time.

The operating companies are largely staffed by their own recruits, but all shareholders second people to the companies. These people usually come for a limited period, say three years, and occupy some of the key managerial and technical positions. They represent about 10 per cent of the total staff numbers.

[7] Shareholders in GASCO are ADNOC (68 per cent), Shell (15 per cent), Total (15 per cent) and Partex (2 per cent).
[8] Law No. 1 of 1988.

Other oil producing companies in Abu Dhabi

Oil was discovered by three companies which took concessions in Marine areas relinquished by ADMA (ADNOC/government do not hold a share in these ventures).

- Total Abu Al Bukhoosh, operated by majority owner Total.
- Mubarraz, operated by owner, a Japanese group.
- Arzanah, previously operated by majority owner Amerada Hess. The field has declined and been handed over to ADNOC.

The total production from these fields never exceeded 100,000 bpd, and has since declined considerably.

In common with Saudi Arabia, Kuwait and Iraq, the main Abu Dhabi oilfields are giant fields providing relatively cheap oil, and therefore high revenues per barrel to the government. The quality of Abu Dhabi oil is good. Approximately 60 per cent of the crude oil exports go to Japan, about 20 per cent to other Asian countries, while the rest goes to a variety of destinations. The LNG is sold to Japan on long-term contracts; some few cargoes have been sold in Europe and elsewhere.

Production of oil and gas has not been interrupted by regional wars, in the case of Abu Dhabi. Looking to the future, market conditions and production costs may be more pressing considerations than politics, but, as this chapter illustrates, Abu Dhabi has been affected by and has developed responses to political imperatives as well as economic ones.

4 Policy imperatives in Washington

Gary G. Sick

During the course of 1997, it became apparent that US policy towards the Gulf was ripe for reappraisal. In the first half of 1998, the Clinton administration took some important steps to modify its declaratory policy towards Iran and to rationalize its military policy. These changes were tentative, however, and could be reversed in response to developments in the region. The US government, for example, showed no inclination to alter its policy towards Iraq and made no effort to challenge its critics in the Congress on the key issues of sanctions towards Iran. Key decisions remained to be taken, and the shape of future US policy in the Gulf was uncertain, but the signs of a major shift away from the 'dual containment' policy seemed clear.

Dual containment had been plagued by serious dilemmas in dealing with Iraq, Iran and the GCC. As indicated by its name, the centrepiece of US strategy had been to contain or to destabilize two regimes in the Gulf: Iran and Iraq. Although this policy attracted widespread but shallow domestic support, it offended a number of important US friends and allies, and over time the benefits of the policy became increasingly difficult to justify in terms of its costs. Implementation has required a large and expensive US military presence and security structure in the Gulf, which has in turn created problems for the very Arab allies it was intended to protect. By the end of 1997 the United States faced a set of changed circumstances in Iran, Iraq and the Arab Gulf states that seemed to presage some possibly painful adjustments over the coming years.

The dual containment policy

The US dual containment policy was adopted by the Clinton administration only three months after it took office when, as noted in Chapter 1, the clearest exposition of what it entailed was provided by then National Security Council member Martin Indyk.[1] In a major departure from past US policy in the Gulf, this formula specifically renounced a strategy of balancing Iran against Iraq. Instead, it operated on the premise that the United States was strong enough to deal with both of these states without the need to pit one against the other.

The policy also spelled out broad guidelines for containing and isolating Iran and Iraq. In the case of Iraq, it centred on maintaining the multilateral sanctions imposed on Iraq after the 1990–91 Gulf war. In the case of Iran, the policy objective was to persuade the Islamic Republic to change its behaviour in a number of areas – terrorism, subversion, violent opposition to the Arab–Israeli peace process, and the pursuit of weapons of mass destruction. To that end, and in the absence of cooperation from its allies, the United States developed an array of unilateral sanctions, including a secondary boycott against foreign companies investing more than $20m in the Iranian oil and gas sector, the Iran and Libya Sanctions Act (ILSA) of 1996, otherwise known as the D'Amato Act after the senator who sponsored it.

Sanctions, in fact, became the US foreign policy instrument of choice in dealing with a range of problems around the world. Between 1993 and 1997, the United States invoked economic sanctions on sixty-one different occasions, more than half the number of sanctions measures imposed over the previous eighty years, and several sweeping new sanctions bills came before the US Congress in 1998. Some of these sanctions were enacted through the United Nations or in cooperation with allies; most, however, were unilateral in nature. A presidential commission estimated the direct cost of economic sanctions to the US economy in 1995 at $15–19bn in lost export sales and up to 250,000 jobs.[2]

The dual containment policy did not specify a military strategy. However, it was evident that the policy would rely primarily on a significant

[1] Martin Indyk, 'The Clinton Administration's Approach to the Middle East', Keynote Address to the Soref Symposium on 'Challenges to US Interests in the Middle East: Obstacles and Opportunities', *Proceedings of the Washington Institute for Near East Policy*, 18–19 May 1993, pp. 1–8. Martin Indyk at the time of this speech had just joined the National Security Council staff. He later became the US ambassador to Israel, and as of this writing he is the Assistant Secretary of State for Near East Affairs.

[2] Under-Secretary of State for Economic Affairs Stuart Eizenstat, testimony before the House of Representatives Ways and Means Trade Sub-committee, 23 October 1997, citing the findings of the President's Export Council.

US military presence in the Gulf region in order to maintain stability and to guard against any aggressive actions by either Iran or Iraq. In practice, this took the form of several squadrons of US aircraft, either flying from bases in Saudi Arabia or else operating on a rotating basis, using facilities in various Arab Gulf countries; deployment of carrier task forces into the region on a frequent but not permanent basis; the deployment of other naval and amphibious forces on a rotating basis; occasional deployments of US ground forces on an exercise basis; and the establishment of large pre-positioned equipment depots, primarily in Kuwait and Qatar, to reduce the transit time for a major insertion of US military forces if and when required.

Security agreements – formal and informal – were established between the United States and the GCC states, and frequent military exercises became routine.[3] This force remained heavily dependent on the infrastructure of the GCC states – airfields, harbours, roads, etc. – as well as their continuing financial subventions.

This substantial and flexible military force provided the necessary core capability to respond immediately to any military threat from either Iran or Iraq, with the additional capability of rapid augmentation from Europe and/or the United States if the threat expanded or was protracted. The dual containment policy explicitly renounced the traditional strategy of balancing the two major regional powers, Iran and Iraq, against each other. On the contrary, it was based on the assumption that the United States was the predominant power in the Gulf with the 'means to counter both the Iraqi and Iranian regimes'.[4] Thus a substantial military presence was inherent in the policy from the beginning.

The Iraqi dilemma

On Iraq, the United States signalled its intention to maintain sanctions as long as President Saddam Hussein was in power. This policy was opposed by a number of states inside and outside the region, but such opposition remained moot so long as Iraq had not complied with UN resolutions calling for the elimination of all weapons of mass destruction.

[3] Britain and France also maintain a smaller but fully coordinated military presence.
[4] Indyk, 'The Clinton Administration's Approach to the Middle East', p. 4.

The United States launched missile attacks against Iraq on a number of occasions in the 1990s, after the 1991 war, with only limited support from other states. There has been some erosion of the original coalition against Iraq, as evidenced by an October 1997 vote in the United Nations Security Council in which the allies split on the question of possibly increasing sanctions against Iraq in retaliation for continued concealment of information and sites involving suspected weapons of mass destruction.[5] The following February, when the United States declared it was prepared to use force against Iraq if necessary to oblige it to cooperate with weapons inspections, and US forces in the area were increased and readied for attack, only the British supported the US threat. However, Security Council members all affirmed the need for the inspections to be completed before sanctions could be lifted.

There is very little debate in the United States about the desirability – even the necessity – of maintaining strict sanctions against the government of Saddam Hussein. There is considerable sympathy in the United States for the plight of the people of Iraq, but Washington has insisted that the misery of the Iraqi people is directly attributable to Saddam, not only because of his earlier aggressive actions but because he has consistently resisted terms for providing humanitarian relief, using the deprivation of his people as a bargaining chip.[6]

With only a few exceptions, the US position has prevailed. However, companies in France, China and Russia – all permanent members of the Security Council – have signed contracts with Iraq for the development of major oilfields that will become active once sanctions have been removed. Several US Arab allies in the Gulf, including the United Arab Emirates and Qatar, have also made clear their opposition to indefinite extensions of sanctions. As also noted in Chapter 6 on Iraq, Sheikh Zayed bin Sultan Al Nahayan, President of the UAE, has been particularly outspoken in calling for greater humanitarian assistance to Iraq and for a closing of ranks among the Arabs, including Iraq, in an effort to put the 1990–91 Gulf war behind them.[7]

[5] On 23 October 1997, France, Russia, China, Egypt and Kenya abstained on a UNSC resolution – the first such case since the establishment of the sanctions regime against Iraq in 1990.

[6] United Nations Special Commissioner Rolf Ekeus estimated that Iraq deliberately deprived itself of $100bn in revenues by refusing to cooperate with UN weapons inspections. See 'For the Record', *The Washington Post,* 31 January 1997.

[7] Quoted in *Reuters*, 1 November 1997.

If and when the UN Special Commission (UNSCOM) determines that Iraq has complied with UN demands for the elimination of weapons of mass destruction, pressure will become intense to remove the sanctions on Iraq's oil sales. The United States has vowed to resist this pressure, and there is considerable likelihood of a confrontation among the coalition members who fought the 1991 Gulf war. The timing of this showdown – like many other issues in the region – is in the hands of Saddam Hussein. If he chooses to cooperate more actively with the UN inspectors, he can advance the date when the allies face a significant difference of views over the sanctions. However, if he chooses to resist the inspections at every turn, he can postpone that moment for a considerable period of time. His decision on 29 October 1997 to bar American members of the UN inspection team was but one example of the kind of defiance Iraq has persistently demonstrated. As suggested in Chapter 6, even after the agreement brokered by UN Secretary General Kofi Annan in February 1998, there has been little reason to conclude that Iraq will readily abandon its strategy of concealing what it can of its capabilities in weapons of mass destruction.

The US dilemma here is clear. Its policy of maintaining sanctions on Iraq has been sustained successfully for more than seven years, but it has failed to bring about the removal of President Saddam Hussein. In the meantime, pressure has been growing in the region and in the West for a modification or lessening of the sanctions policy, primarily for humanitarian reasons. If the United States perseveres in its determination to maintain sanctions until Saddam Hussein is gone, it is very likely that a confrontation among the allies will occur by the turn of the century.

The Iranian dilemma

The original formulation of dual containment, as described by Martin Indyk in 1993, called for changes in Iran's behaviour in five areas: (1) support of international terrorism; (2) support for Hamas and attempted sabotage of the Arab–Israeli peace talks; (3) subversion through support of Islamic movements in Sudan and elsewhere; (4) acquisition of conventional weapons that would permit Iran to dominate the Gulf region; and (5) prospective acquisition of weapons of mass destruction.

In pursuance of this policy, Indyk said, the United States would attempt to persuade European powers, Japan, Russia and China to reject Iranian requests for weapons of mass destruction and conventional weapons that might constitute a regional threat. The United States would also oppose development loans to Iran by the World Bank and the IMF, and would seek to persuade its allies to maintain pressure on Iran so it could not 'pursue normal commercial relations'. Although some of his original formulations were resisted by the Department of State, Indyk's description of the policy proved to be entirely accurate. As of mid-1998, the United States had in place an array of executive orders, regulations and statutes which not only prohibited any economic interaction between the United States and Iran, but also threatened to impose sanctions on companies of other countries that invested in the Iranian oil and gas sector.

The US dual containment strategy has indisputably created political and economic problems for Iran. Over the past several years, Iran has experienced difficulties in attracting investment capital for its ailing oil industry. Although some companies may have stayed away from Iran simply because the terms of the deals were not particularly attractive, there is no doubt that the threat of US sanctions enhanced commercial caution. Of eleven offers that Iran made prior to 1998, only two gained support; at the same time, a series of offers by Venezuela with scarcely better terms was snapped up.

The challenge to US sanctions has come in three areas. First, Turkey contracted with Turkmenistan for construction of a gas pipeline across Iran in a deal that was announced almost immediately after the passage of ILSA. The State Department responded with a tentative ruling that this did not require sanctions since it was to the benefit of the parties at either end of the pipeline, with only incidental benefits to Iran. Second, the Clinton administration was sensitive to the problems facing the new republics of the former Soviet Union in Central Asia and the Caucasus in finding export routes for their oil and gas. This problem was acknowledged in the initial executive orders that prohibited oil purchases from Iran but allowed oil swaps.

More significant than either of these was the announcement in September 1997 by the French oil major Total, together with state-owned partners Gazprom of Russia and Petronas of Malaysia, that they had concluded a $2bn deal to develop an Iranian gas field. The US administration reacted cautiously, launching an investigation of the contracts and generally resisting any rush to sanctions. After dragging out the investigation for some months, the administration finally announced that ILSA would be waived in this instance on national security grounds. Imposing sanctions would almost certainly have sparked a confrontation with several US allies, probably including the European Union; failure to impose sanctions, however, may have opened the floodgates for many other companies to begin dealing with Iran. Certainly, US companies have been chafing at the constraints still operative on them, while foreign companies increasingly pursue lucrative prospects in Iran.[8]

The dual containment policy with respect to Iran attracted increasing criticism from many different quarters. During 1997, a growing chorus of former senior US policy-makers began to call for changes, arguing that the sanctions policy had little effect on Iran's behaviour, and proposing a more subtle programme of carrots and sticks. The US business community, which had remained totally silent when sanctions were being voted, began to organize itself in opposition to further closing off of US international business opportunities, in an effort spearheaded by a movement called USA Engage, which questioned the principle of the use of sanctions as a foreign policy tool by Congress. Meanwhile, the international community became increasingly vocal in its opposition to US extraterritorial efforts to impose its policies.

An apparent watershed came in January 1998, with the interview given by Iranian President Mohammed Khatami to CNN, in which he paid tribute to the Western civilization represented in America and called for dialogue at the cultural and academic level. As elaborated by Edmund Herzig in Chapter 7 on Iran, the changes taking place in the Islamic Republic, manifest since Khatami was elected in May 1997, are potentially far-reaching and their importance has not been lost on members of the Clinton administration. One of the first signals sent by Washington was the designation of the Iranian dissident group Mojahedin-e-Khalq as a terrorist organization. In June 1998 US Secretary of State Madeleine Albright made a ground-breaking speech in which she signalled a readiness in the administration to consider a normalization of relations and called for the formulation of a 'road map' for achieving this.

[8] Robert Corzine, 'Iran Declines to Keep Oil Deals for US', *Financial Times*, 2 July 1998.

President Clinton followed with a conciliatory message on the occasion of the World Cup football match between Iran and the United States in France. Iran won the match, which did no harm to relations. President Khatami reacted cautiously to the official overtures, however, calling for deeds, not just words.[9]

Even accepting these clear signals of willingness on the part of the US administration to redirect its policy towards Iran, the process will not be easy in view of the structure of statutes and regulations that will continue to act as legal impediments and irritants in relations between the United States, Iran and key US allies. Moreover, it would be a mistake to assume that Iran will necessarily be a willing participant in any US effort to improve relations. In some respects, the effect of the US sanctions has been almost exactly the opposite of that intended. The Iranian economy, instead of collapsing under the pressure of sanctions, has done rather well, at least until the oil price collapse of 1998. Iran has not yet fully recovered from the huge dislocations caused by the revolution, but the trends are more positive than negative.[10]

As also subsequently fully discussed by Dr Herzig, a process is under way whereby Iran is engaged in a major diplomatic effort to restore relations with the Arab states, beginning with its immediate neighbours in the Gulf. Initial success was manifest at the summit of the Islamic Conference Organisation (OIC), hosted by Tehran in December 1997, where there was high-level representation from member states, notably Saudi Arabia. Iran will serve as the chair of the OIC until the turn of the century.

One of the original purposes of the dual containment policy was to support the Arab–Israeli peace process by severely limiting the ability of its two most important critics to interfere. It is therefore especially ironic that Iran's efforts to establish itself as a leader of the Islamic world have been greatly enhanced by the policies of Prime Minister Netanyahu of Israel. The near collapse of the Arab–Israeli peace process has validated Iran's rejectionist position in Iranian and many Arab eyes.[11]

[9] 'Khatami Calls on U.S. for Deeds of Goodwill', *International Herald Tribune*, 2 July 1998.

[10] Jahangir Amouzegar, 'Adjusting to Sanctions', *Foreign Affairs*, Vol. 76, No. 3, May/June 1997, pp. 31–41.

[11] In private, Iranian officials say that they have been greatly encouraged by their contacts with the Arab states, who tell them that Iran has been proved right about the bankruptcy of the peace process and about US and Israeli intransigence.

Because of these factors and the domestic opposition in Iran likely to greet any precipitous opening to the United States, Tehran will probably be more cautious about seeking a rapprochement with Washington than it might have been a few years earlier. In short, Iran and the United States will not find it easy to extricate themselves from the web of acrimony that they have woven so tightly over the past generation, even if they wish to do so. It is more likely that such a process will proceed gradually, with a series of small steps, rather than a sudden breakthrough or policy reversal.

The GCC dilemma

Relations between the United States and the GCC states have generally remained strong – economically, politically and militarily. The shock of Saddam Hussein's invasion of Kuwait forged the closest relationship between the United States and the GCC states in the long history of US–Arab cooperation in the Gulf, including formal and informal military ties. The emergence of the United States as the major military power in the region, together with continuing concerns about potential Iraqi and Iranian threats, led to a degree of military and political cooperation that would have been difficult to imagine only a decade earlier.

However, as time passed, this relationship also began to show signs of strain. The Arab states of the Gulf have grown uncomfortable with the continued sanctions against Iraq, which they perceive as unfairly punishing the civilian population of that fellow Arab country. The GCC states have also begun to develop more normal relations with Iran and have tended to be increasingly sceptical of assertions from some US quarters about the direct Iranian threat to GCC stability. In both cases, the Arab states have been intensely aware of the fact that they must find ways to live with their two larger and more powerful neighbours and that the unrelenting hostility of the US dual containment policy could backfire against their own interests.

These concerns have been magnified by internal opposition within the GCC states to the prominent US military role in the region. Bombings in Riyadh and Dhahran dramatized the degree of opposition to the US military presence. Governments have also been stung by opposition group charges that it is illegitimate for Islamic governments to rely on the 'infidels' of the West to protect their core Islamic values. Those attitudes have been reinforced by the long-standing Arab nationalist critique of dependence on the West. Opponents of the Western presence have become progressively more outspoken in calling for the removal of US forces. Increasingly, it has become popular to view US actions as a means of selling ever greater quantities of military equipment to the Arabs while leaving them in a state of dependency.

These attitudes have been strengthened by opposition to the Arab–Israeli peace process and objections to the policies of Israeli Prime Minister Netanyahu. As the peace process has faltered and as the Arab states have grown increasingly critical of US support for Israel, it has become more acceptable to speak openly and critically of US policies. Arab rulers must be concerned that too close an embrace of the United States under these circumstances can undercut their own legitimacy and subject them to attacks both from the Islamists and from the Arab nationalist intellectuals in their own societies. The clearest example of this trend is the stout resistance of most of the Arab states to participation in the MENA (Middle East, North Africa) economic conference in Doha in November 1997, despite intensive and high-level diplomatic efforts by the United States to persuade them to attend.

There are also internal stresses within the GCC states deriving from the difficulties of meeting growing economic demands from their own populations at a time when oil revenues are flat or declining. These pressures are gradual, and there is no immediate sense of crisis; however, the pressures are growing each year as a new generation of citizens, better educated and in many cases lacking meaningful jobs, begin to contemplate their role in the future of these countries. Political challenges to the existing order seem likely in the future, although the nature and timing of such challenges are not predictable.

As of 1998, these changes had not seriously affected the underlying relationship with the United States. The GCC states are uncomfortably aware of their own military weakness and lack of unity. They realize that they are unable to defend themselves against a serious thrust from either Iraq or Iran, and after two regional wars between 1980 and 1991, they understand that this is a possibility that cannot lightly be ignored. Moreover, some of the smaller states of the GCC may quietly take comfort from the US presence as a cushion of protection against their larger Arab brothers and neighbours.

The dilemma for US policy – and for America's Arab allies in the Gulf – is how to balance the need for US protection and support with the growing uneasiness about dependence on the West. In the US case, this has taken the form of reducing the visibility of US forces in the region by moving them to remote desert areas while relying heavily on naval forces rather than shore-based forces wherever possible. In a sense, this seems to be a gradual return to the 'over the horizon' strategy of years past. The 'horizon' came much closer as a result of the two Gulf wars, and it could do so once again if there is another direct military challenge. But as memories of the Iraqi invasion of Kuwait fade, pressures will no doubt continue to reduce the appearance of reliance on the United States and to seek political and possibly military arrangements on a regional basis that would provide at least a partial alternative to exclusive dependence on the United States and the West. These pressures are likely to become much more serious if and when the first two dilemmas of the US dual containment policy come to a head.

Conclusions

After more than five years, the US dual containment policy is approaching a turning point. Within the United States there is powerful political support for the policy, but questions are increasingly being raised by senior foreign policy figures inside and outside the administration as well as major business interests. Those differences are being exacerbated by pressures from the international community, opposing unilateral US sanctions and actions. Actions of the Clinton administration with respect to Iran in 1998 appear to signal a trend away from full-fledged containment of Iran, at least.

In the view of many observers, including this author, the dual containment policy was designed to take advantage of an anomalous set of circumstances produced by the 1990–91 Gulf war: US military and political predominance; Arab fears overcoming traditional aversion to perceived dependence on outside powers; and the extraordinary weakness of both Iraq and Iran as the result of a decade of war and revolution. Those circumstances are now beginning to change perceptibly, and the original policy is being called into question.

As the twentieth century draws to a close, the United States appears to be exploring cautiously a new security policy for the Gulf region that would fall somewhere between the extreme dependence on allies and regional surrogates that characterized the period before the Iranian revolution, and the exuberant unilateralism of the period following the second Gulf war. These early signals appear to reflect the judgment that future access to oil is more likely to be driven by technologies and markets than by politics, that the predations of Saddam Hussein are unlikely to be repeated in the near future, and that the threats emanating from the region could potentially be managed more effectively in a regional context rather than by exclusive reliance on the United States as the political and military hegemon. The United States seems certain to remain the major power in the region, but it is possible to imagine a reduction of the US military presence and a concomitant increase of attention to diplomacy and regional security arrangements.

All these judgments are tentative and could be upset by the kind of surprises that have occurred all too frequently in the Gulf over the past half-century. Four flash points remain that could – and at some point probably will – interfere with a smooth transition to a more regionally based strategy:

- A breakdown in the Arab–Israeli peace process could spill over into the Gulf politically or even militarily, as it has in the past.
- Either the reinvigoration or the collapse of Saddam Hussein's regime in Iraq could create a vortex of instability in the heart of the Middle East.
- A new turn of the Iranian revolution could shred the delicate fabric of goodwill that has been created under President Khatami.
- A political or social breakdown in one of the key Arab states of the region could shatter the illusion of contented stability in the Gulf Cooperation Council states.

In any of these cases, the United States would again be called upon to play a central role in the affairs of the region. It would again have to choose – as it has so many times in the past fifty years – between an indirect proxy role and some form of military intervention, either independently or as the leader of a coalition.

History provides no reliable guidance. It is fair to say, however, based on the above analysis, that the existing policy relationships are not going to remain static. As the twenty-first century begins, the constellation of political, economic and military relationships in the region is likely to look very different from the way it looked in the mid-1990s. It is a cliché, but nevertheless true, that the only certainty is change.

5 Europe and the United States: conflicting approaches

Rosemary Hollis

There are very distinct differences between the United States and its European allies over how to respond to developments in the Gulf region. They differ in capabilities, analysis and approach. Although as a whole the European Union represents a larger economic player than the United States in the post-Cold War world, its members do not all act as one and the EU cannot back common foreign policy initiatives, even when these are forthcoming, with military force projection. There are cultural differences too. In the United States domestic politics is very often a powerful determinant of foreign policy and there is a tradition of imbuing US conduct abroad with moral purpose. In Europe, by contrast, foreign policy is more blatantly linked to the pursuit of national economic interests and there is less regard for normative considerations.

Thus, in the light of US championship of a 'new world order' in the 1990s based on the designation of 'rogue regimes' and a policy of containment towards the states concerned, European governments have frequently balked at demands by the United States for allied support for its hegemonic lead. Such reticence has been particularly apparent in the Gulf since the end of the 1991 war. On Iraq, France has led European critics of the US line that sanctions cannot be lifted unless and until the government of President Saddam Hussein disappears. On Iran, for much of the 1990s, Germany has been in the forefront of a European drive to influence the Islamic Republic through 'critical dialogue' rather than isolation, and the Europeans have been united in their opposition on principle to US secondary boycott legislation. With respect to the GCC states, meanwhile, the Europeans are fierce competitors with the United States for commercial markets and defence sales.

European imperatives

Oil is the single most important European interest in the Gulf region. Europe is more dependent than the United States on oil imports from the area, not least because, after the North Sea and the Mediterranean, the Gulf and potentially the Transcaucasus are the nearest sources of supplies for European consumers. Admittedly, as discussed in Chapter 2, the late 1990s have witnessed a superabundance of oil supplies on the world market, which is likely to defer a substantially increased reliance on Gulf reserves in the near to medium term. However, given the proximity, magnitude and relative accessibility of Gulf oil for Europe, the long-term prospect is for increased dependency.

Against this background, the implications of the US posture in the Gulf during the 1990s have weighed somewhat heavily with strategic planners in the European energy industry. As set out in preceding chapters, the US policy of dual containment involves the maintenance of multilateral sanctions on Iraq; unilateral sanctions on Iran; a forward military presence in the Gulf (and Turkey); and defence agreements with the GCC states. One of the consequences of these arrangements, presumably incidental rather than planned, has been a distortion of the Gulf oil market. Iraq was kept off the market altogether (except for supplies to Jordan and smuggling) until implementation of the 'oil-for-food' deals with the United Nations began in 1996. Investment in the Iraqi oil sector has been prevented since 1990. Meanwhile, passage of the 1996 Iran-Libya Sanctions Act (ILSA) forced would-be investors in its energy sector to risk penalties in the United States if they went ahead.

From the point of view of European consumers, the absence of Iraqi supplies for an extended period and the curtailment of the Iranian potential have not meant hardship, since plenty of oil has been available from GCC producers and other sources. However, with some notable exceptions, GCC countries have not welcomed foreign investment in their oil production, at least not on terms that were attractive to major energy companies. One consequence, as discussed previously, has been heightened interest in more remote fields elsewhere. Yet the attraction of alternative locations could diminish, if more opportunities for equity sharing open up in the Gulf. In the meantime, European oil companies have been chafing at the obstacles preventing them from developing a stake in the most extensive and accessible resources in the world.

Iraq, Iran and international energy companies are not the only ones to lose out as a result of a set of circumstances at least partially attributable to US Gulf policy. At the strategic level, security of access to Gulf energy supplies has become dependent on a structure largely designed and maintained by the United States. As discussed by Gary Sick, this gives cause for concern among Washington's Gulf allies. By the same token, there is no room for complacency among energy-consumer countries, not least in Europe. If, for some reason, the United States were to abandon its military posture in the Gulf, the security system would collapse. If, however, the United States overplays its hand, there could be a backlash even among its Arab allies in the region. A crisis disrupting supplies from one or more of the major GCC producers would have immediate and potentially serious repercussions for the energy market as a whole, since neither Iran nor Iraq, even if allowed, could marshal additional capacity quickly. Potentially, a breakdown in the security arrangements underpinned by the United States could disrupt Gulf supplies across the board.

In broad terms, therefore, European energy interests in the Gulf have become beholden to US policy. Under dual containment it has not been possible for Europeans or others to pursue and develop multiple sources of energy supply from the region, in either the immediate or the medium term. This is the background against which European policy differences with Washington may be understood.

From the perspective of the United States, however, the Europeans appear to be taking advantage of the US commitment to Gulf security, and the force deployments and economic penalties for US companies that this entails. In so far as the Europeans take for granted the American defence shield and pursue their own commercial interests in the area, Washington, and in particular the US Congress, has made deliberate efforts to curtail or punish what it sees as 'freeloaders'. ILSA was prompted not only by a drive to increase the pressure on the Iranian regime, but also by a determination to prevent US allies from stealing opportunities which US companies had already been barred from pursuing. However, by resorting to a secondary boycott, the United States incurred open denunciation from all its European allies and risked initiating a trade war.

The transatlantic debate

The single most important European objection to US policy towards Iran has been the secondary boycott legislation embodied in ILSA. The Europeans object on principle to the notion that the United States can legislate to penalize foreign companies for their activities in other parts of the world. ILSA has been attacked as a blatant transgression of the principles of international free trade and if the United States had punished Total and its partners Gazprom and Petronas for deciding to invest in the Iranian gas sector in a blatant challenge to the D'Amato legislation, the EU would have been expected to respond by taking up the issue with the World Trade Organization. However as was noted in Chapter 4, the US administration avoided a showdown on that occasion by waiving penalties on grounds of US national interest.

The Europeans also differ from US official policy in their threat assessments. On the issue of Iran's alleged nuclear ambitions, the Europeans have tended to show more sympathy than the Americans for Iran's position. They see Israel's nuclear capability as a provocation to all its neighbours in the region, and would prefer the international community to adopt a more even-handed and comprehensive approach to the question of regional arms control. Perhaps more importantly, the Europeans have not suffered the same affronts at the hands of the Islamic Republic as the Americans did at the outset of the revolution, and they regard Iran as an important player in the region which cannot be sealed off and isolated. Instead, they advocate engagement and dialogue.

On the question of Iran's involvement in terrorism, while granting that Iranians were behind the assassinations of a number of Iranian nationals abroad who opposed the Islamic Republic, European governments have been less exercised than the Americans about Iranian support for Islamic resistance movements opposed to Israel. The European assessment has been that both the Lebanese Hezbollah and Palestinian Hamas are 'home-grown' movements, with their own reasons for attacking Israelis irrespective of Iranian backing. More importantly, members of the EU have questioned the usefulness of a policy of economic boycott as a means to bring about a more cooperative attitude in Tehran, with respect to either the West or Iran's Arab neighbours. Rather, they have claimed, such a policy is more likely to strengthen the hand of hardliners and drive

Tehran to greater belligerence. One of the arguments used by Europeans against the US resort to sanctions affecting the Iranian energy sector has been that the consequences are more likely to be felt by future generations of Iranians than by the targeted regime. In addition, expressing a view recently adopted by US companies, various commentators and some within Congress,[1] Europeans have argued that sanctions should be a policy of last rather than first resort, adopted only when there is sufficient international solidarity to promise effectiveness.

The difference in approach between Europe and the United States was clearly demonstrated in the case of the so-called Mykonos affair.[2] In April 1997 a German court sentenced four men found guilty of murdering Iranian Kurdish dissidents in a shooting at the Mykonos restaurant in Berlin in 1992. In passing judgment the court implicated senior members of the Iranian government. Not only Germany but all the EU members then withdrew their heads of mission from Tehran for consultations. Iran responded in kind, and countrywide demonstrations took place denouncing Germany, along with the United States and Israel. The Iranian parliament urged the government to seek compensation for contracts on which Germany had defaulted, to freeze Iranian investments in East Germany and to prosecute German firms which sold chemical weapons to Iraq in the 1980s. Yet the Iranian leadership, as well as expressing outrage at what it called a politically motivated decision by Germany, talked of the storm soon blowing over. The Europeans were also confident that the problem could be resolved in due course. Apparently misreading the mood in Europe, the United States greeted the Mykonos verdict as a vindication of its stance. However, with the election of Iran's new President Mohammed Khatami in May 1997, the EU governments were even more convinced than before that sanctions were not the appropriate way to deal with Iran. By the end of the year a formula had been worked out between the EU and Tehran for the return of their ambassadors to post.

During the first half of 1998, European relations with Iran did indeed recover. In June the Italian Prime Minister visited the Islamic Republic, becoming the first EU premier to make such a trip since the Greek Prime Minister visited in 1992. The event was taken as a signal of a readiness in

[1] 'Senate Forms Task Force To Examine Sanctions Policy', *MEES*, 6 July 1998; Richard Haas, 'Sanctions Almost Never Work', *The Wall Street Journal*, 23 June 1998.

[2] Saeed Barzan, 'Iran: Bitter Fruits of Mykonos', *Middle East International*, No. 548, 18 April 1997, p. 12.

the EU, spearheaded by Italy, to further economic ties. Italy's trade with Iran was valued at $2.3bn in 1997 and the Iranians greeted the Italian initiative with enthusiasm.[3]

Undoubtedly, European policy has been influenced by its own commercial as well as energy interests. Ten per cent or so of European oil imports come from Iran.[4] In 1996 EU imports from the Islamic Republic were valued at $5.3bn and Iranian exports to Europe at $6.3bn.[5] By contrast, the United States has ceased all trade with Iran. Europeans also have debts outstanding in the Islamic Republic which inevitably translate into an interest in the general health of the Iranian economy. From a business standpoint, Iran represents a major growth opportunity, a big market, a strategically placed conduit to Central Asia. What has held Iran back from capitalizing on foreign commercial interest has been a combination of factors, only partially attributable to the US sanctions. A poor regulatory environment and uncertainties surrounding Iranian economic policy have also played a part. However, in 1998 the Iranian energy sector sought to overcome some of the obstacles by offering a range of new investment opportunities including, for the first time, onshore.[6]

Meanwhile, as discussed by Edmund Herzig in Chapter 7, the Islamic Republic might even be depicted as a stabilizing influence in the region overall, or at least in Central Asia and the Caucasus. As explained by Dr Herzig, one of the by-products of the US containment policy towards Iran has been the advent of a deliberate policy in Tehran to reach out to its neighbours and avoid isolation. By 1997, Iran was seeking ways to improve relations with the Arab Gulf states too, and progress was particularly evident with Saudi Arabia.

By 1998 it could be said that the European approach to Iran appeared to have been vindicated, as Tehran and Washington took their own steps to reduce the tensions between them. That, at least, would be the European interpretation, but in the United States the advocates of containment remained cautious and would certainly not concede that their policy had failed. On the contrary, they were still arguing that a tough line was the best way to bring about change in Tehran. Assistant Secretary of State

[3] 'Iran Says Prodi's Visit Signals Thaw With EU', *International Herald Tribune*, 1 July 1998.
[4] *BP Statistical Review of World Energy*, 1997.
[5] IMF, *Direction of Trade Statistics*, March 1998.
[6] Robert Corzine and Robin Allen, 'Tehran Reopens Doors to Vast Oil and Gas Resources', and Robert Corzine, 'Top Companies Lured by Huge Opportunity', *Financial Times*, 1 July 1998.

Marc Grossman, meanwhile, warned against allowing Iran to become the conduit for Caspian energy exports on the grounds that: 'Iran is not only risky as a route for energy; it also keeps control of the region's energy reserves in fewer hands.'[7]

Grossman's words echo the logic which galvanized the US-led coalition into going to war to eject Iraq from Kuwait and prevent it from extending its reach to the oil-rich eastern province of Saudi Arabia in 1990–91. At that time, the Europeans were in agreement with the United States that it would threaten all their interests if a single regime were to monopolize control of the majority of the Gulf's oil reserves. However, while the Americans and the British have remained convinced that Iraq still poses a threat, some eight years later the analysis of the situation in other European capitals has evolved differently.

The French lead the way in Europe in cautioning against sustaining sanctions on Iraq indefinitely. From their perspective, the continuation of the sanctions threatens to invite a new danger of collapse and chaos inside Iraq itself, creating a kind of vortex of insecurity at the head of the Gulf. Certainly, there is concern that Iraq has been so weakened by the cumulative effects of sanctions that it may not be able to play a stabilizing role in the future and its population could emerge from the humiliating and debilitating experiences of the 1990s in an angry and vengeful mood, as implied by Raad Alkadiri in Chapter 6. With this possibility in the wings, France and others of like mind have argued that sanctions should only be applied in relation to the specific goal outlined in UN Security Council Resolution 687, namely the elimination of Iraq's WMD capabilities and its long-range missiles.

According to the French it is essential that the letter and spirit of the resolution be adhered to scrupulously. Sanctions must remain until Iraq is given a clean bill of health by the UN weapons inspectors.[8] However, the moment that this happens, it would be illegal to keep the oil embargo in place. As also mentioned by Gary Sick in Chapter 4, the French are at one with the Russians and Chinese in arguing that only by sticking to this interpretation of the situation can they hold out an incentive to the Iraqi regime to comply. Their view attained central place in the agreement

[7] See Stephen Fidler, 'US "Should Not Exclude Iran from Oil Deals"', *Financial Times*, 9 July 1998.
[8] Press statement made by French Foreign Minister Hervé de Charette, following his meeting with Iraqi Deputy Prime Minister Tariq Aziz, Paris, 14 January 1997, *SAC/97/14-Iraq*.

worked out between UN Secretary General Kofi Annan and the Iraqi government in February 1998. Iraq was essentially promised the reward of an end to trade sanctions, aside from defence-related equipment, if it cooperated fully in the speedy conclusion of the UN investigations of its nuclear, chemical and biological weapons programmes.

In the crisis which prompted Annan's mediation with Baghdad, France was among those opposing the use of force against Iraq. Even so, like Kofi Annan himself, the Europeans generally acknowledged that the military threat mounted by the United States and supported by Britain helped to put pressure on the Iraqi regime. In effect, it was tantamount to a coordinated approach by the United States, its European allies and Russia, involving a combination of threats and incentives which appeared to work. However, if, as Raad Alkadiri suggests (Chapter 6), Iraq will never reveal everything about its weapons capabilities, there will come a point when the UN inspectors can do no more. At that juncture, the French and some other Europeans will argue in favour of closing the files and moving on, for fear of the continued debilitation of Iraq, while the United States and probably Britain are likely to argue for an extension both of sanctions and of the 'oil-for-food' formula instead.

As is frequently pointed out in Washington, France has financial and oil interests at stake in Iraq which must colour its perspective. French companies have signed up to develop Iraqi energy production, along with Russian and other firms, as soon as the embargo is lifted. Only then will France and other creditors stand a chance of recouping some of their outstanding debts. Obviously these are compelling considerations. However, they are reflective of the broader context within which the issue of how to deal with Iraq is viewed outside the United States. Underlying the views of Europeans and others on Iraq is, once again, a concern about their long-term oil interests.

Of all the Gulf oil producers, Iraq represents a special case for Europe, since supplies from there can be transported directly to the Mediterranean, via pipelines across Turkey and/or Syria. In this way, supplies from Iraq may be cheaper to access and could be brought out irrespective of whether the United States is ensuring safe passage through the Gulf waters or not. That said, however, Iraq is in some ways less immune than its Gulf neighbours to the dangers which could spill over as a result of renewed confrontation in the Arab–Israeli sector.

Conclusion

The prospect of collapse of the Arab–Israeli peace process loomed large in the summer of 1998. For Europe, given its proximity and economic exposure to, and energy dependence on, the Middle East and North Africa regions overall, this has been cause for alarm. Yet the United States has traditionally sought to keep the Europeans at a distance from Arab–Israeli peacemaking. Since the end of the 1991 Gulf war, US predominance across the region has been unrivalled. Consequently, not only in the Arab–Israeli sector but also in the Gulf, the Europeans must operate on the assumption of US hegemony. This dictates a policy of persuasion rather than confrontation. However, a level of increased acrimony between Washington and its European allies can be anticipated over how to deal with a recalcitrant government in Israel, as well as one in Baghdad. As of this writing, though, it seems that, at least with respect to Iran, there are signs of a growing convergence of views.

6 Iraq under sanctions: diminishing returns

Raad Alkadiri

The last decade of the twentieth century has been a dark era for Iraq. This country, with its abundant oil wealth, good agricultural land, the makings of an industrial infrastructure and a pool of highly educated and skilled labour, had the potential to take a leading role in regional economic development. Instead it has gone into deep economic decline, the middle class has been virtually wiped out, the population is suffering from impoverishment and malnutrition, and the regime of President Saddam Hussein – which has endured through the war with Iran in the 1980s and the 1990–91 Gulf war – is in survival mode, in the face of internationally supported attempts to unseat it.

Since its invasion of Kuwait in August 1990, Iraq has been subjected to the most extensive and rigidly enforced international sanctions regime in modern history. The passage of over thirty United Nations resolutions over a period of eight years, along with a variety of other independent government decisions, has placed sweeping restrictions on Iraqi imports and exports. These measures have also weakened the sovereignty of the state: Western-imposed air-exclusion zones in the north and south of the country prohibit Iraqi military aircraft from entering nearly two-thirds of Iraq's airspace; the country's military-industrial complex has been constrained; and the United Nations has approved and implemented long-term monitoring programmes which limit Iraq's future weapons capability and block the import of all but a selection of 'humanitarian' goods.

In spite of their severity, sanctions – and more specifically the economic embargo – have proved only partially effective. As this chapter will show, even when considered within the narrow terms of UN Security Council Resolution 687 (the 'ceasefire' resolution passed at the end of the war in 1991), the embargo has failed to encourage Baghdad's complete cooperation with international weapons inspectors – a fact borne out by the crises in Iraq's relations with the Security Council since late 1997. Sanctions have proved even less effective when judged according to Washington's broader interpretation of their purpose, namely to induce the overthrow of the current Iraqi regime. Their main impact, in fact, has not been political, but rather social and humanitarian in nature, as large sections of the Iraqi population have been forced to endure devastating socio-economic decline.

This imbalance between the political and the economic impact of the embargo has generated misgivings about the sanctions policy among many governments, including key US allies. Nevertheless, Washington remains unflinching in its commitment to maintaining economic pressure on Baghdad. The resulting strains upon the cohesion of the coalition that forced Iraq out of Kuwait and the difficulties in maintaining a unified policy have potentially serious implications for future efforts to contain Iraq. Moreover, the effectiveness of US policy in serving its longer-term interests in the Gulf is in doubt.

Socio-economic collapse

The international community's stringent enforcement of the embargo has effectively undermined the foundations of the Iraqi economy. By stopping oil exports, except, latterly, under UN supervision, it has deprived the state of its principal source of finance. Estimates of the contribution of oil revenue to the overall economy prior to the 1990–91 Gulf war vary: according to the Economist Intelligence Unit (EIU), oil sales accounted for 60 per cent of GDP in 1989 and 75 per cent in 1990 when calculated at the market exchange rate; using the official exchange rate, the Iraqi government put the figure at 15 per cent for 1990. Irrespective of the exact figure, it is clear that the contribution of oil exports to the economy was immense. Moreover, oil sales accounted for over 90 per cent of Iraq's foreign currency earnings during this period.[1]

[1] EIU figures quoted in Peter Boone, Haris Gazdar and Athar Hussein, 'Sanctions Against Iraq: Costs of Failure', report prepared for the Center for Economic and Social Rights, 1996, p. 8; Center for Economic and Social Rights, 'Unsanctioned Suffering: a Human Rights Assessment of United Nations Sanctions on Iraq', report published May 1996, p. 6.

Macroeconomic deterioration

The macroeconomic consequences of this loss of revenue – estimated at over $130bn to the beginning of 1998 – have been severe. Combined with the infrastructural devastation inflicted during the Gulf war, it has led to a major contraction of the economy. Determining the precise extent of this economic decline is difficult owing to the absence of data – Baghdad continues to regard economic indicators as a matter of national security. Nevertheless, a variety of estimates exist. The Food Aid Organization (FAO), for example, suggests that real GDP declined by around 75 per cent in 1991, while the OECD calculated in 1995 that nominal GDP had shrunk by 35 per cent since August 1990.[2] Meanwhile, real per capita income fell from US$4,000 in 1980 to $485 in 1993[3] and estimates for 1997 put the figure at around $200, well below 1950 levels.[4]

The period since the imposition of sanctions has also witnessed extremely high levels of inflation. Some estimates suggest the annual inflation rate exceeded 300 per cent between 1991 and 1995. While the figure subsequently dropped, it remained in the region of 100–200 per cent in early 1998. Two factors account for the postwar rise in the index. First, the decline in oil revenue forced Iraq to reduce its import levels, which fell from $10.3bn in 1988 to a mere $400m in 1991, and the Security Council's decision in 1992 to freeze Iraqi bank accounts abroad added to the import financing problems experienced by the government. The fall in imports, and a concomitant decline of domestic production, have resulted in a shortage of basic tradeable goods, including food items. Almost 70 per cent of cereal requirements and 75–80 per cent of all calories consumed were imported prior to sanctions. Although the government has encouraged greater agricultural production, this has been insufficient to overcome the food deficit, in part due to a lack of machinery, high-grade seeds and fertilizer. The consequence has been a major increase in the relative prices of these goods.

The second factor fuelling inflation was the government's postwar

[2] Food Aid Organization/World Food Programme, 'FAO/WFP Food Supply and Nutrition Assessment Mission to Iraq', report published 3 October 1997, p. 3.

[3] Abbas Alnasrawi, *The Iraqi Economy: Oil, Development, Wars, Destruction and Prospects, 1950–2010,* Westport, Greenwood Press, 1994, p. 168.

[4] Boone, Gazdar and Hussein, 'Sanctions Against Iraq', p. 13.

pursuit of an expansionary monetary policy in a bid to fund its budget deficit. The state's fiscal shortfall was brought on in part as a result of its attempts to preserve public-sector employment and wage levels (which were increased repeatedly, from September 1991 to the end of 1995). The government also intervened regularly in the black market to maintain the value of the dinar. However, its concurrent policy of printing money merely fuelled the decline of the currency, thereby increasing inflationary pressures. The official exchange rate, which has remained at its pre-war level of 0.311 dinars to the dollar, bears no resemblance to the real value of the currency which, judged by the black market, began a steep decline in the aftermath of the imposition of sanctions. From 8 dinars per dollar in mid-1991, it fell to over 100 dinars per dollar by early 1993. The plunge in the currency was hastened drastically in 1995 when the government, faced with a reported deficit of $600m, intervened heavily on the black market, leading to increased demand for dollars and a fall in the value of the dinar to almost 3,000 per dollar in December that year.[5]

In a bid to reverse the dinar's rapid decline, the government implemented tough monetary and fiscal measures. In addition to slashing the rate at which domestic banks traded the dinar to 1,000 against the dollar (state banks have been permitted to buy and sell hard currency outside the official rate since February 1994), the state froze the printing of money; except for projects deemed to be in the national interest, individual ministries have had to obtain direct presidential approval for additional expenditure. Public-sector salaries were also frozen in December 1995, and a ban on new recruitment implemented. Parallel to this, subsidies on the cost of utilities and state services were reduced, and the scope and collection of direct and indirect taxes were increased. Other schemes to reduce liquidity included the sale of state assets, such as property and small factories, and a ban on the initiation of loss-making projects.

Nevertheless, the impact of these fiscal and monetary measures appears to have been limited. Fluctuations in the value of the dinar have continued, linked primarily to external factors. Baghdad's announcement in January 1996 of its willingness to negotiate with the United Nations on plans to allow limited oil exports under the terms of UN Security Council Resolution 986, and the subsequent implementation of the agreement in

[5] 'Iraq Explains Sanctions-busting Plans', *Financial Times*, 9–10 November 1996.

December of the same year, both led to a short-term strengthening of the currency: for example, the value of the dinar went up from 1,740 per dollar in November 1997 to 850 the following month. However, the trend has generally been downwards: although some stabilization of the currency occurred in early 1997, at around 1,250 dinars per dollar, with the arrival of the first major shipment of the oil-for-food deal, ensuing political crises and bureaucratic delays in authorizing food and medical import contracts subsequently undermined the dinar, which had fallen to almost 1,700 to the dollar by January 1998.[6]

Social conditions

The rise in the relative price of basic goods, including food items, and the parallel collapse in the value of the dinar, have led to a major deterioration in standards of living in Iraq, particularly among the middle class and civil servants. With the massive decline in real wages (the average public-sector wage has dropped to an estimated $3–5 per month in some areas), food access has emerged as a major problem. Despite efforts to generate additional income, the majority of the Iraqi population suffers from reduced, if not insufficient, food intake.[7] Indeed, a World Food Programme representative warned in late 1995 that more than 4m Iraqis were at severe nutritional risk, including 2.4m children under five and 600,000 pregnant or nursing women.[8] Malnutrition among children has led to significant increases in the levels of associated health problems, including stunting, wasting and disorders such as kwashiorkor (the incidence of which increased from 2 per 100,000 in 1990 to 51 per 100,000 in 1994) and marasmus (14 per 100,000 to 465 per 100,000). A highly effective government rationing system has removed the threat of widespread famine. However, rations, which supplied around 50 per cent of pre-war daily calorific intake when introduced in September 1990,

[6] Food Aid Organization/World Food Programme, 'FAO/WFP Food Supply and Nutrition Assessment Mission to Iraq', p. 3.

[7] Dieter Hannusch, 'Quick Reference Paper on WFP Assessment Mission to Iraq', report published 25 September 1995, p. 2. The WFP report estimated the market price of subsistence foods for a family of five to be about $26 per month.

[8] 'Time Running Out for Iraqi Children', *The Food Aid Organization of the United Nations System News Update*, 26 September 1995.

were cut in September 1994, reducing the provision to a mere 34 per cent of pre-war intake.[9] Moreover, owing to their cereal base, the rations are deficient in a number of minerals and vitamins (notably A and C), as well as in animal protein.

A deterioration in healthcare conditions has also been witnessed. Prior to sanctions, 90 per cent of the population had access to primary healthcare, part of the extensive range of welfare services provided by the state. However, the healthcare system has increasingly decayed, primarily because of import restrictions on key equipment and spare parts.[10] This includes a shortage of anaesthetics, surgical instruments and disposable equipment. Supplies of basic drugs, including antibiotics, have also been insufficient: drug imports fell from $360m worth per year prior to the Gulf war to an estimated $13m in 1996, while limited pre-war domestic production has effectively ceased.[11] As a result of these factors, hospitals and health centres have been unable to respond adequately to increased levels of disease. Poor sanitary conditions – associated with impaired water and sewage systems, and the lack of basic supplies – have led to significant increases in incidences of infectious, parasitic and waterborne diseases, including typhoid, cholera and malaria (an epidemic occurred in 1992–4). Independent observers also report increased cases of diseases such as diabetes and cancer.

The impact of these factors is seen throughout the population. Mortality rates reported by hospitals have increased significantly, despite a decline in attendance. Children have proved particularly vulnerable: between 1990 and 1995, child mortality rates rose five-fold, with death rates among those aged under five reaching 12.8 per 100 live births according to the Iraqi Ministry of Health. Independent reports suggested that infant mortality doubled over the same period, reaching 16.1 per 100 live births.[12]

[9] World Health Organization, 'The Health Conditions of the Population in Iraq Since the Gulf Crisis', *WHO/EHA/96.1*, March 1996, p. 2.
[10] See, for example, 'Iraqi Health System Close to Collapse Says *WHO* Director-General', *Press Release WHO/16*, 27 February 1997.
[11] Center for Economic and Social Rights, 'Unsanctioned Suffering', p. 21.
[12] 'Time Running Out for Iraqi Children'; World Health Organization, 'The Health Conditions of the Population in Iraq Since the Gulf Crisis', p. 5.

Oil-for-food

Faced with the prospect of socio-economic collapse under the continued sanctions regime, both the international community and the Iraqi government have been forced to react. International concern over the humanitarian consequences of the economic embargo was a prime motivation behind UN Resolution 986, which permitted Iraq oil sales worth $2bn every six months to fund the import of food and medicine. Baghdad initially rejected the oil-for-food deal, arguing that its strict terms for the purchase and distribution of humanitarian goods represented an infringement on Iraq's sovereignty (the government rejected on similar grounds two earlier, more stringent versions of the deal, authorized by the UN Security Council in August and September 1991). Senior Iraqi officials, foremost among whom was President Saddam Hussein, also feared – with some justification – that Resolution 986 was conceived as a means of deflecting humanitarian pressure for a lifting of the economic embargo. Arguments proposed by some Iraqi technocrats that the oil-for-food process could ultimately weaken the sanctions regime by facilitating the re-establishment of links with foreign oil companies were rejected. However, the accelerated decline in the value of the dinar in late 1995, combined with pressure from Russia and France (both of which emphasized that they did not support an imminent lifting of the embargo), forced Baghdad to reverse its position and announce in January 1996 its willingness to negotiate the implementation of the limited oil sales formula.[13]

The arrangements that were eventually worked out went into effect in December 1996. An initial prognosis from the EIU that the economy would expand 35 per cent in 1997 appears to have been exaggerated. The domestic food supply situation did improve after implementation began, but the commensurate increase in the level of official rations proved subject to reverses, as Baghdad sought to gain UN approval for new distribution procedures and to overcome import impediments. Delays in receiving approval for individual deals from the UN Sanctions Committee on Iraq (Committee 661) hampered the flow of imports: by the beginning of August 1997, Iraq had imported only 52 per cent of planned

[13] 'Iraq: Sanctions Stratagems', *Oxford Analytica Daily Brief*, 10 October 1995; 'Iraq: Oil-for-Food Formula', *Oxford Analytica Daily Brief*, 22 January 1996.

food deliveries for the first six-month phase of the accord (December 1996–June 1997).[14] The second phase faced even more delays owing to its late implementation.

In effect, the original oil-for-food deal offered no solution to Iraq's fundamental socio-economic problems, even in the short term. It was purposely limited in scope, designed to meet the minimum requirements of the Iraqi population while maintaining the economic stranglehold imposed on the regime. As it turned out, the $1bn (after deductions) which the government received every six months for food and medical goods fell well short of the amount which international organizations estimated Baghdad would need to meet its humanitarian requirements. A UN estimate put the total annual cost of food and medicine imports needed by Iraq at $4bn; the FAO estimated that the cost of imported food requirements alone would total $3.106bn per year. Consequently, the organization concluded in its October 1997 report that existing ration levels remained insufficient to eliminate under-nutrition throughout the country.[15] Meanwhile, the import cost estimates did not account for the expenditure required to restore the healthcare system, agricultural production and, eventually, Iraq's economic infrastructure.

By early 1998 it was broadly agreed that an expanded formula would be needed. The United Nations gave the go-ahead for this in February 1998, more than doubling the value of oil that Iraq could export and allowing for a wider range of imports for 'humanitarian' purposes. Even so, the bureaucratic procedures involved in operating the scheme remain cumbersome. Strict application of the rules banning 'dual-use' equipment (that is, equipment which could be used in the production of weapons) means that anything from fertilizers to spare parts for machinery can be blocked. Meanwhile, as the government of Iraq has submitted to the United Nations, it must be allowed to repair and upgrade its oil installations if it is to manage to export the value of oil permitted under UN supervision. This point was conceded by the United Nations in spring 1998, but by the following July, Iraq was complaining that clearance for the shipment of specific items of equipment was being blocked.

[14] WFP Emergency Report No. 32, 8 August 1997.

[15] Center for Economic and Social Rights, 'Unsanctioned Suffering', p. 3; Food Aid Organization/World Food Programme, 'FAO/WFP Food Supply and Nutrition Assessment Mission to Iraq', p. 12.

Policy determinants

Iraq's acceptance of the oil-for-food formula is testimony to the direct impact of sanctions on policy, but not necessarily in the sense intended by the proponents of the trade embargo. Overcoming the impact of the embargo and lobbying for its removal have been principal focuses of Iraqi policy since 1991. However, as borne out by the confrontations between Baghdad and the United Nations over weapons inspections in late 1997 and February 1998, the economic stranglehold imposed on Iraq has only partially succeeded in forcing Saddam Hussein's regime to comply with the terms of Security Council resolutions or to modify its political approach. Iraqi officials emphasized, not disingenuously, that the crises of late 1997 and early 1998 related to their disillusion and frustration with US policy, and did not reflect Iraq's unwillingness to comply with UN terms. Nevertheless, it remains the case that from 1991 onwards Baghdad has only begrudgingly cooperated with the UN's task of eliminating Iraq's non-conventional weapons and its long-range missile capability, in spite of the fact that, according to Resolution 687, satisfying its criteria is the key to the lifting of sanctions.

Baghdad's obstructionism, however, has not been total or unremitting. Cooperation between the Iraqi authorities and the International Atomic Energy Agency (IAEA) led to relatively rapid progress on the nuclear front. The majority of components and facilities of what was a well-funded programme for the development of indigenous nuclear weapons were discovered and destroyed by the end of 1992. Subsequent IAEA activity has focused on follow-up monitoring and searching for concealed proscribed equipment and materials. Reporting to the UN Security Council in October 1997, the agency suggested that while it had no absolute assurance that readily concealable items did not exist, its investigations were reaching the point of diminishing returns.

Yet, driven by their respective political considerations, neither Baghdad nor the staunchest advocates of the sanctions regime chose to highlight the agency's conclusions publicly. For the Iraqis at that time, it would have undermined their argument that they were receiving no recognition for cooperating with international weapons inspectors; for Iraq's opponents, it flew in the face of allegations against Baghdad for non-compliance with UN Security Council resolutions. In 1998, however, the issue has been addressed more openly in the Security Council, with Russia in the forefront of those urging that the nuclear file be closed first, even if other issues remain outstanding.

Iraqi dealings with the UN Special Commission (UNSCOM) that is charged with responsibility for eliminating biological and chemical weapons programmes have, by contrast, been characterized by repeated concealment of data and facilities, and intermittent confrontation. In practical terms, this policy has resulted in Iraqi vacillation between cooperation with and obstruction of the UNSCOM mission. The shift between these two approaches has been determined by a number of factors related both to the Iraqi leadership's perception of the likelihood of a removal of the economic embargo and to international pressure (whether in the form of recommendations from states regarded as more 'sympathetic' to Iraq's plight, or threats and punishment, including military action). Non-cooperation and hostility on the part of Iraqi officials were particularly evident in the period immediately following the Gulf war ceasefire. However, the perpetuation of sanctions led to a partial reversal of Iraqi policy in September 1993. This included Baghdad's belated acceptance of the long-term monitoring and verification programme approved in late 1991 by the UN Security Council in Resolution 715.[16] While this more positive approach has generally endured, intermittent and well-publicized temporary breakdowns in the relationship have nevertheless continued, with periodic confrontations which reached crisis proportions in February 1998 when the UN Secretary General Kofi Annan had to intervene personally. Iraq has also levelled charges of political bias against some commission members, including former commission executive chairman Rolf Ekeus and his successor Richard Butler.

In part, Iraqi obduracy has reflected Baghdad's frustration and anger with the durability and the impact of the embargo, and the absence of a definite timetable for its removal – sentiments which drove Iraq to mass troops on the Kuwaiti border in October 1994 and which underlie its adoption of a more confrontational approach to UN inspectors, particularly those from the United States, since late 1997. However, officials

[16] On this period, see Raymond A. Zilinskas, 'UNSCOM and the UNSCOM Experience in Iraq', *Politics and the Life Sciences*, Vol. 14, August 1995, pp. 230–31.

have also impeded UNSCOM inspections for a variety of other reasons, including claims that access to certain facilities threatened national security and infringed on Iraqi sovereignty. While these latter charges undoubtedly reflect genuine Iraqi concerns, they also form part of the government's systematic policy to conceal the extent of its WMD programmes.[17] The extent of Baghdad's efforts to hide information was dramatically exposed following the defection in August 1995 of Hussein Kamel, President Saddam Hussein's son-in-law and the former head of Iraq's military industrialization programme. In a bid to limit the potential damage of Kamel's revelations to UNSCOM – and equally to undermine his value to Western governments – Iraq was forced to provide the commission with previously concealed information, at a time when it appeared that UNSCOM was close to concluding its investigations and demolition work in certain areas. This new information, which included over 500,000 pages of data, revealed, among other things, the indigenous production of missile engines (Project 1728), a larger and more sophisticated chemical weapons programme than hitherto suspected and the first real indications of the extent of Iraq's offensive biological weapons programme.[18] Baghdad also then committed itself to full and indefinite cooperation with UNSCOM.

The Iraqi government's failure to abide consistently by this pledge led to the possibility of further sanctions being imposed. Nevertheless, as the October 1997 UNSCOM report to the Security Council indicated, renewed efforts by the commission since 1995 have ultimately resulted in considerable progress towards uncovering and neutralizing Iraq's ballistic missile and chemical weapons capabilities. For example, all but two of the 819 long-range missiles imported by Iraq up to 1988 have been accounted for, and new elements of Iraq's VX nerve agent programme have been identified. However, as the report also made clear, Iraq had persisted in its attempts to conceal areas of its non-conventional weapons capacity, particularly its biological weapons programme, despite Baghdad's

[17] Iraqi government claims to have destroyed unilaterally – and in contravention of Resolution 687 – components, equipment and documents relating to its prohibited weapons programmes are also suspected of forming part of this policy and have considerably complicated the task of weapons inspectors.

[18] United Nations, Security Council, S/1995/864, 11 October 1995. Iraq later claimed that Kamel was responsible for hiding the subsequently revealed information on Iraq's weapons programme from UNSCOM and the IAEA (p. 8).

claims to have complied in full with UN requirements.[19] A closed-door briefing by Richard Butler to the UN Security Council in June 1998 reportedly addressed his concerns that the Iraqi government was still not complying fully with UNSCOM investigations.[20]

Baghdad's perpetual pursuit of this secretive and intermittently confrontational policy clearly runs counter to its objective of securing the full or even partial removal of the economic embargo. Iraqi duplicity has reinforced distrust of the integrity of Saddam Hussein's regime and its disclosure claims, thereby extending the weapons verification and investigation periods. Moreover, it has hampered attempts by 'sympathetic' states, notably France and Russia, to push for a lifting of the blockade.

Nevertheless, Iraq's behaviour is not devoid of political logic. Baghdad's policy is consistent with the leadership's refusal to submit to the role of the vanquished since the Gulf war. Consequently, it has been unwilling to acknowledge through its actions the retributive spirit underlying the imposition of sanctions. In practical terms, this is manifest in Iraq's repeated attempts to negotiate the terms of Resolution 687 and its anger at the Security Council's failure to reward progress made towards eliminating its non-conventional weapons. Iraqi defiance in the face of what it portrays as a Western, and more specifically US, plot to bring the country to its knees also serves a domestic political purpose, deflecting popular criticism generated by the deteriorating economic and social conditions away from the regime. It remains a matter of conjecture whether Iraqi officials subscribe to this conspiracy theory, although anecdotal evidence suggests that some may.

However, while such political factors are undoubtedly important, the main consideration guiding Iraqi policy appears to be strategic. The progress made by UNSCOM and the IAEA up to 1998 has undoubtedly weakened Iraq's current and future military capacity, and potentially weakened its capacity to resume a major regional role. Decision-makers in Baghdad believe that the total elimination of the country's weapons of mass destruction and the capacity for their delivery poses a significant medium- and long-term threat to national security, particularly given the

[19] United Nations, Security Council, S/1997/774, 6 October 1997. On Iraq's biological weapons capabilities, see also Raymond A. Zilinskas, 'Iraq's Biological Weapons: the Past as Future?', *Journal of the American Medical Association*, Vol. 278, No. 5, pp. 418–24.
[20] Michael Littlejohns, 'UN Studies Secret Data on Iraq Arms', *Financial Times*, 4 June 1998.

presence in Iraq's immediate strategic environment of hostile states, such as Iran and Israel, which possess a non-conventional weapons capability. Under these circumstances, and given the paramount political objective of regime survival, Baghdad has clung to its policy of partial compliance with UNSCOM, irrespective of the economic consequences, in the apparent hope that the sanctions regime will either collapse or be eased significantly before Iraq's full WMD capability is uncovered.

Regime stability

If sanctions have been only partially successful in forcing Iraq to cooperate with the United Nations, they have proved an even less effective means of achieving the broader political purpose ascribed to them by a number of powers, notably the United States. Unlike France, Russia and a number of Arab states, which have adopted a narrow view of the embargo's intent, arguing that it should be lifted when Iraq meets in full the terms of Resolution 687, Washington, backed by London, has viewed sanctions as part of broader efforts (including covert operations in the north of Iraq)[21] aimed at containing and ultimately inducing the overthrow of the present regime. US officials have repeatedly expressed this objective in veiled terms by emphasizing that Iraq must achieve what Washington clearly believes is beyond the present Iraqi regime, namely compliance with all UN Security Council resolutions – including those relating to peaceful intent and respect for minority rights – before sanctions can be removed. However, in a clear expression of intent, in 1997 US Secretary of State Madeleine Albright indicated that Washington's support for the embargo would be sustained while Saddam Hussein's regime remained in power. US officials have also hinted publicly that they would be willing to consider longer-term economic assistance for a new administration in Baghdad.[22]

[21] See, for example, Jim Hoagland, 'Humiliated in Iraq, Again', *Washington Post*, 8 May 1997; Jim Hoagland, 'How the CIA's Secret War on Saddam Collapsed: a Retired Intelligence Operative Surfaces With Details and Critique of US Campaign', *Washington Post*, 26 June 1997.

[22] Speech by Madeleine Albright, 26 March 1997, reproduced in *Middle East Economic Survey*, Vol. 40, No. 13, 31 March 1997, pp. D1–D4; 'American Official Discusses the Desire to Talk to a New Iraqi Leadership', *The Iraq File*, No. 66, June 1997, p. 54.

That this unbending approach has been effective in containing Baghdad and limiting Iraq's regional role has been cause for some satisfaction with US policy in Washington. As of this writing, however, the United States has failed to accomplish its broader objective. President Saddam Hussein appears to be firmly entrenched in power, and certainly in a stronger position in the late 1990s than he was in the immediate aftermath of the Gulf war. The regime has gradually restored and extended its authority throughout much of the country since the low point of February 1991, when it faced simultaneous popular uprisings by the Kurds in the north and the Shi'a in the south and briefly lost control of fourteen of Iraq's eighteen provinces. The southern rebellion was crushed rapidly, if brutally, by the end of March 1991. There have been persistent reports since of violence aimed at government forces; however, this has occurred on a limited scale and, overall, order has been maintained, in part as a result of the repression of actual and potential opposition by military and security forces.[23]

In the north, Baghdad has benefited from the inability of Kurdish leaders to reach a long-term political compromise that would allow them to take advantage of the effective autonomy extended to the three northern governorates as a result of the air exclusion zones imposed by US and allied forces. The political honeymoon enjoyed by the Kurds between 1991 and 1994, during which time local elections took place and a quasi-provincial government emerged, was shattered when long-standing rivalries and disputes over economic control between the two main Kurdish parties led to armed conflict. The continuation of fighting despite Western mediation efforts ultimately provided the central government with an opportunity to re-establish a security presence when, in August 1996, it intervened in support of Kurdish Democratic Party forces in their operation to capture Irbil from the control of the rival Patriotic Union of Kurdistan.[24] However, restoration of full political control over the region has eluded Baghdad, largely owing to Kurdish distrust of the Iraqi leadership and a measure of international surveillance.

The regime's task of asserting and maintaining its authority in the central and southern regions of Iraq has in many ways been assisted by

[23] This latter point has been emphasized by UN Human Rights investigators. See, for example, United Nations, Economic and Social Council, *E/CN.4/1997/57*, p. 4.
[24] For details see United Nations, Economic and Social Council, *A/51/496/Add.1*, pp. 3–8.

the embargo. In the face of ever-increasing economic hardship, the focus of the vast majority of the Iraqi population has been on survival. Nevertheless, the regime has faced organized domestic opposition, as substantiated since 1991 by regular reports of executions, particularly within the military. In June 1996, a plot involving mostly junior and middle-ranking officers – allegedly organized with CIA support – was uncovered.[25] Reports of a major coup attempt in August 1997, this time allegedly involving figures linked to the regime, have also surfaced, although their accuracy has not been verified by any official sources.[26] At the same time, regime figures have been the target of attacks, most spectacularly the assassination attempt against the president's elder son, Uday, in December 1996. In response to these incidents, in 1997 the president reportedly approved a plan to tighten security and increase the protection provided for officials throughout the country.[27]

In spite of the apparent deterioration of internal security, challenges to the regime's survival have been largely contained. Opponents have been eliminated, or cowed by the threat of violent repression. At the same time, organizing effective opposition action has been severely complicated by the existence of an all-pervasive security apparatus, composed of numerous and often competing agencies led by close and trusted associates of the president. Infiltration by the security services at all levels of the military, combined with the constant rotation of officers between commands, has undermined the army's ability to mount an effective challenge to the present order, in spite of the fact that antipathy to the regime's policies is reportedly extensive within the armed forces, including the Republican Guard. Moreover, the apparent unity of aim has been insufficient to overcome differences in opinion within the military over how to bring about political change. Military opponents of the regime are also concerned about the lack of a unified and powerful external opposition, which they perceive as necessary to provide political support for an army-led coup.[28]

[25] See 'News About the Thwarting of an Overthrow Attempt in Iraq', *The Iraq File,* No. 56, August 1996, p. 39.

[26] Nayel Mukheiber, 'The Secret Behind the Sudden Moves and Changes in Baghdad: the Story of the Most Dangerous Overthrow Plan', *The Iraq File*, No. 70, October 1997.

[27] 'Saddam Approves Plan to Tighten Security, Aimed at Confronting the Violations and Assassination Attempts', *al-Quds al-Arabi*, 7 March 1997.

[28] Interview with Major-General Mustafa Mahdi al-Salahi, 13 June 1996, reprinted in *The Iraqi File*, No. 55, July 1996, p. 31.

The survival of the Iraqi regime has been further assisted to some extent by the defection of an estimated 100 or more senior political, diplomatic and military figures. While many of those who fled Iraq subsequently joined externally based opposition groups and attempted to maintain links with sympathizers within the country, their departure has removed from the regime's midst potential sources of threats to its existence. Nevertheless, this exodus does suggest that ruptures are occurring in the regime's support base. At the same time, there has been repeated speculation about growing rifts and power struggles within the ruling inner circle, which gained increased credence after the 1995 defection of Hussein Kamel and the shooting of Watban Ibrahim al-Hassan, the president's half-brother and a former Minister of the Interior. Recent conjecture, angrily denied by Iraqi officials, has focused in particular on alleged feuds and confrontations between the president's sons and his three half-brothers.[29]

Although these confrontations have allegedly been linked to battles for control of Iraq's limited unofficial commercial activity, it is not immediately evident that any existing rifts are a direct result of the pressure imposed on the regime by the economic embargo. Strains of this type between senior figures within the ruling elite were also witnessed prior to the sanctions era. Moreover, opposition and Western reports of the extent of infighting, and the consequent threat it poses to regime stability, are most probably exaggerated. Members of the regime recognize that they would be the principal targets for violent popular retribution should the regime collapse. Nevertheless, there are indications that the circle of trusted family members and associates around the president may be narrowing. Perhaps more threatening are suggestions of developing rivalries between key security organizations, which, if correct, could ultimately undermine the mainstay of the regime. Reports that Saddam has been considering a reorganization of the military and security high command structure, linking all elements under the direct control of his younger son Qusay,[30] may reflect the president's desire to impose tighter and more centralized control over the security services in order to counteract these rivalries.

[29] See, for example, 'Saddam Arranges for "Family Reconciliation", Urges His Two Sons to Settle Differences with Their Uncles', *al-Quds al-Arabi*, 11 March 1997; 'Baghdad Denies an Internal Struggle and Annan Expects an Extension of Work on 986', *al-Quds al-Arabi*, 31 January 1997.
[30] Sean Boyne, 'Qusay Considers a Reshuffle for Iraq's Command Structure', *Jane's Intelligence Review*, September 1997, p. 416.

The regime's attempts to establish a broader support base since 1995 have lent weight to speculation about presidential concerns regarding threats to its clan and security-based power structure. Efforts to rekindle the role of the tribes within the country, which have been a feature of domestic politics since the 1980s,[31] were accelerated following Hussein Kamel's defection, focusing particularly on prominent Sunni tribal groups such as the Shammar, Dhafir and al-Dulaim. Early reports suggested that the government would review the succession of agricultural reform measures introduced since 1959, thereby restoring land appropriated by the government to its former tribal owners. Moreover, it was claimed that the regime was considering the gradual reintroduction of legislation first established by British mandate authorities in the 1920s which gave tribal leaders authority to collect taxes, and determine social and legal issues related to their kinsfolk.[32] However, this speculation has been overplayed, as the regime's increased enthusiasm for tribalism proved short-lived. Indeed, Baghdad subsequently introduced measures clearly designed to limit tribal authority and negate any idea that it might be willing to acquiesce in a dual system of government within Iraq. In March 1997, Iraq's highest executive body, the Revolutionary Command Council (RCC), passed decree no. 24 outlawing tribal traditions as grounds for disobeying state laws. The decree insisted that, while the tribes constituted a genuine part of Iraq's social and national fabric, tribalism should not be viewed as supra-legal. Moreover, it emphasized that national interests should assume precedence over tribal ones.[33]

The aftermath of the Hussein Kamel defection also witnessed symbolic attempts to revitalize the role of the Ba'ath Party, which diminished inexorably during the 1980s and early 1990s. However, this has done little to revive the party's institutional or ideological independence, which has been systematically subjugated to the demands of regime politics. Thus, in spite of rhetoric to the contrary, the party effectively operates as but one of a number of vehicles of policy implementation.[34]

[31] See, for example, Amatzia Baram, 'Neo-tribalism in Iraq: Saddam Hussein's Tribal Policies 1991–1996', *International Journal of Middle East Studies*, Vol. 29, 1997, pp. 1–31.
[32] Emir Tahiri, 'Saddam Hussein Tries to Revive a Tribal Order After the Collapse of Party Control', *al-Sharq al-Awsat*, 15 May 1996.
[33] Adeed Dawisha, 'Iraqi Politics: the Past and Present as Context for the Future', in John Calabrese (ed.), *The Future of Iraq*, Washington, DC, The Middle East Institute, 1997, p. 14.
[34] See for example, discussion with former presidential adviser Sadeq Sha'aban, *The Iraq File*, December 1996.

Changes to various levels of party cadres in mid-1997, with the retirement and replacement of long-standing officials, do not signal any imminent change in the party's current function. Rather, these moves form part of the now almost routine reshuffles within the organs of the state, including the council of ministers, designed in part to keep officials on their toes and prohibit the emergence of alternative bases of power. The elevation of a new generation of officials may serve an additional purpose of invigorating the state's capacity to resist the challenges it faces as a result of its present predicament.

Thus far, Baghdad's efforts to expand its support base have not extended to genuine political reform. Senior officials have hinted at regular intervals that gradual changes in the structure of government are planned, including moves towards the establishment of civil institutions in place of the present revolutionary ones. In December 1996, for example, Iraqi Foreign Minister Mohammed Saaed al-Sahaf reportedly told pro-government Iraqi groups in Jordan that the formation of a 'civilian' council of ministers, composed of technocrats of known liberal and pro-reform persuasion, was likely in the near future.[35] However, these statements have remained mere rhetoric. They appear to be intended largely as a tactical ploy, designed to generate some external momentum towards a lifting of sanctions. Various sources, both internal and external, have made it clear to Baghdad that the absence of reform is contributing to the longevity of the embargo. This has reportedly been the word from the French government, as well as from King Hussein, who regards political liberalization as fundamental if Jordanian mediation efforts on Iraq's behalf in Washington (initiated, it appears, at Baghdad's request) are to stand any chance of success.

If persistent media reports are to be believed, the Iraqi president was given the same message by his half-brother, Barzan al-Tikriti, who in a memorandum of early 1997 proposed sweeping political changes, including the abolition of the RCC and the one-party system.[36] However, the Iraqi president's reportedly cool response to the latter proposals – he

[35] 'In a News Report from Amman: Baghdad Consents to Major Changes to Accompany the Period After Oil and Food: Saddam Prepares to Form a Technocratic Government Including Kurdish and Liberal Personalities', *The Iraq File*, No. 61, January 1997. See also interview with Speaker of the Iraqi National Assembly Saadoun Hammadi, *The Iraq File*, No. 66, June 1997.

[36] Saddam Hussein himself suggested similar reforms in the run-up to the October 1995 referendum on his presidency.

postponed dealing with the suggestions, probably indefinitely, by turning them over instead for consideration by a special committee, composed of close associates who are regarded as hard-line[37] – suggests that no major reforms are to be expected. Sanctions, it appears, have done little to reverse the regime's unwillingness to countenance any form of political opposition, whether real or imagined. With the challenges posed by the deteriorating domestic security situation and by the strains on the existing power structure remaining manageable, the regime seems unlikely to introduce measures that would undermine its exclusive rule, and potentially threaten its survival.

The sanctions dilemma

The failure of the embargo to overturn the political status quo in Baghdad has done nothing to diminish Washington's commitment to maintaining UN sanctions on Iraq. If anything, the US administration has become even more firmly wedded to the use of sanctions since the bruising experience of February 1998, when Washington stood virtually alone in threatening military action against Iraq to make it cooperate with UNSCOM inspections. Yet the economic blockade is not merely of questionable value as a policy tool for a US administration unwilling to engage in more direct means to achieve its objectives. It is also counterproductive to Washington's interim and long-term regional objectives.

The Clinton administration's inflexibility, bolstered by an implacable Congress, poses problems for its policy in a number of ways. Its most obvious impact has been to weaken further the already fragile allied coalition. Washington's continued hard line towards Baghdad, despite the clear disproportionality between the humanitarian and political impacts of the embargo, has exacerbated pre-existing fault lines in the Security Council caused by differences over policy objectives in Iraq. This in turn has undermined the unity of purpose which has been key to the success of

[37] One of the committee's members, RCC Vice-Chairman Izzet Ibrahim, reportedly criticized Barzan and Iraq's UN Ambassador Nizar Hamdoun for their 'Western-inspired' views during the meeting of the Regional Command of the Ba'ath Party in October 1996. For a partial text of the memorandum, see *al-Quds al-Arabi*, 8 April 1997; 'Saddam Has Not Rejected Barzan's Memorandum, He Has Formed a Political Committee to Study It', *al-Quds al-Arabi*, 21 March 1997.

the UN's efforts thus far. Both France and Russia have been indirectly critical of the United States through repeated expressions of concern over the humanitarian toll which sanctions have taken on the Iraqi people. They have also argued that Baghdad has been given insufficient credit by the Security Council for its progress thus far in eliminating its non-conventional weapons. Furthermore, the two states have opposed the use of force against Iraq: both were critical of US military strikes against targets in southern Iraq in September 1996 (France also refused to participate in monitoring the extension to the no-fly zone over the area), and neither state was willing to support US-led military action to resolve the weapons inspection crisis when it came to a head in February 1998. On that occasion, the extent of Arab as well as French and Russian opposition to the use of force (especially in the absence of any clear guarantees of what this might achieve) and the emergence of dissenting voices even on the US domestic front patently gave the administration pause. In the event, the Clinton administration appears to have been relieved to accept the deal worked out with Baghdad by Kofi Annan to defuse the crisis. However, the upshot is that Washington has little left in the way of a policy for containing Iraq and the ambitions of its leadership except the continuation of sanctions.

Meanwhile, the mediation efforts of the UN Secretary General were effective in winning over Baghdad because the resulting deal held out the promise that UN inspections could be concluded within a reasonable timeframe, given renewed Iraqi cooperation, raising expectations that the end could be in sight for the economic sanctions. At the same time, there has been a gradual diminution in support for Washington's hard-line position on Iraq. Among the GCC states, even Kuwait showed unease about the potential fallout from resort to force during the confrontation of February 1998 and Saudi Arabia refused to grant Washington permission to use the kingdom as a base for attacks. Latterly, there have been intimations that the Saudis are increasingly uncomfortable with the consequences for Iraq of the continuation of sanctions. Elsewhere in the region, commercial interest, strategic concerns and growing unease about the plight of the Iraqi people have resulted in a more sympathetic policy towards Iraq. A growing popular animus towards Washington provoked by its perceived failure to pursue an even-handed policy in the region, particularly in relation to Israel, has also contributed to this shift.

President of the United Arab Emirates Sheikh Zayed has publicly lamented the continued suffering of the Iraqis and has signalled a growing willingness to accept Iraq's political and economic rehabilitation. Meanwhile, Egypt had called on the UN Security Council to clarify the precise terms which Baghdad is required to fulfil for sanctions to be lifted, ahead of the new understanding reached in February 1998, which appeared to go some way towards this. Even Jordan, which had shifted to a more hard-line approach towards Saddam Hussein's regime after August 1995, has subsequently been more critical of what it perceives as Washington's attempts to interfere in the domestic affairs of Iraq.

The growing divisions within the international community are compatible with Iraq's foreign policy agenda; since 1991 Baghdad has attempted to engender a humanitarian and commercial lobby in support of lifting the economic sanctions. That said, as long as almost all states remain unwilling to defy the embargo to any significant degree, Iraq's ultimate goal also remains unrealized. In the absence of future Iraqi aggression, maintaining this cohesion in the medium-to-long term (which will be critical for the success of UN efforts to monitor Iraq's imports and its compliance with prohibitions against reacquiring non-conventional weapons) appears to depend on a US willingness to be flexible. Despite UNSCOM claims of continued Iraqi recalcitrance, it is not readily apparent, for example, that coalition unity could withstand a US veto of Security Council efforts to ease significantly or lift the blockade, given the growing reservations about the justification and consequences of the embargo. Even if it can, Washington's unwillingness to compromise will necessitate the further expenditure of valuable diplomatic capital on an issue that rates only moderately among its global political priorities.

Viewed in the longer term, Washington's rigidity could still backfire politically even if the broader US purpose of sanctions is realized. The apparent disregard of US policy makers for the humanitarian cost of sanctions has allowed Baghdad to portray itself domestically as the victim of premeditated US malevolence. The expansion of the oil-for-food deal is portrayed as only a mechanism for retaining the overall sanctions regime. In addition to shifting the blame for the suffering of Iraqi people onto Washington, Baghdad benefits from traditional popular antipathy towards the United States (generating a similar sentiment towards those states perceived as backing US policy). Even among those who recognize the Iraqi regime's role in creating the present situation, there is a widespread belief that US policy is designed to perpetuate the status quo in order to maintain a weakened Iraqi state. Under these circumstances, there is no guarantee that an alternative Iraqi regime would necessarily be better disposed towards the United States and its policy objectives in the Middle East than the present one: history suggests that revolutionary administrations in Iraq have tended to adopt a radical nationalist orientation which has been anti-Western in its outlook, driven in part by domestic popular sentiment. Nor is it readily apparent that a new Iraqi administration would necessarily enhance regional political stability. Even a military overthrow – Washington's favoured scenario – will inevitably be followed by some period of violent instability. Moreover, existing political differences within the armed forces are likely to manifest themselves quickly as internal power struggles. Almost any foreseeable outcome would be destabilizing and could potentially encourage indirect or direct intervention by neighbouring states.

Washington's present policy towards Iraq also limits its available contingency options in the event of a significant change in the strategic environment in the Middle East. At the end of his discussion on US policy (Chapter 4), Gary Sick posits a number of regional developments which could interfere with a smooth transition to a more indigenously based security structure. In fact, at present a number of scenarios are conceivable that might make the rehabilitation of Iraq a less unpalatable policy option for the United States. Iranian acquisition of nuclear weapons, for example, would significantly shift the strategic balance in the Gulf. The concern already expressed by Washington and its regional allies about this possibility suggests that, were it to occur, containment of Iran would become an overriding priority for the United States, and one which could force it to abandon 'dual containment' in favour of a return to some form of constructive engagement in a bid to use Iraq as a counter to the regional threat posed by the Islamic Republic. Iraq's importance to Washington would also increase in the event of a serious crisis in Saudi Arabia, particularly if this threatened significant disruption of oil supplies to the advanced industrial world. Despite appearing remote, these or other circumstances (such as an Iranian move to close the Straits of Hormuz) would probably necessitate a resumption of Iraqi oil exports in order to avoid extended panic in the oil markets and an economically damaging rise in the price of crude.

As of late 1998, however, it is difficult to see how, in any of these cases, the United States could make the needed policy shifts as swiftly as circumstances would demand. The 'demonization' of the present Iraqi regime will be difficult to undo, especially in the Congress. Moreover, Baghdad's willingness to ally itself with the United States under the scenarios envisaged cannot be counted on. For the government in Iraq, even the removal of sanctions will not turn it overnight into a friend of the United States and for the time being Washington is the biggest obstacle in the way of attaining that objective.

Conclusion

The incongruity between the socio-economic and the political impact of the UN-imposed embargo on Iraq has become increasingly evident over the past eight years. Economic and welfare conditions within the country have deteriorated to the point of near collapse, bringing poverty, disease and virtual starvation for large proportions of the population. Yet no commensurate political impact has been witnessed. While Iraq has been contained and militarily weakened, its compliance with the demands of Resolution 687 has been gradual, grudging and, thus far, incomplete. Sanctions have also failed to change significantly the policy goals of the regime or to dislodge it from power; indeed, it was in a stronger position seven years on than it was when Resolution 687 was passed.

That the sanctions regime has remained fully in place is due in large measure to Washington's dogged commitment to the policy as a means of changing Iraqi behaviour, and ultimately its political order. However, the embargo is limiting Washington's policy options, and, more importantly, increasingly appears to be in danger of threatening many of the regional objectives that the United States hopes to achieve. To a large extent, this is due to the refusal of the United States to respond to the interests and concerns of its regional and international partners, which has weakened the cohesion of the Gulf war alliance. While no state has broken ranks overtly, this is an ever more likely prospect so long as Washington refuses to formulate policy on a basis acceptable to its international and regional partners. The hazard of doing otherwise is not that Iraq will ultimately escape its political and economic isolation. Rather, it is that the common purpose which has been present within the international community for most of the 1990s will be lost, thereby undermining – possibly fatally – the efficacy of already active monitoring and verification programmes which offer a long-term means of containing Iraq's military potential.

7 Iran and its regional relations

Edmund Herzig

Introduction

The period since the election of President Mohammed Khatami in May 1997 has been one of dynamic change in Iran's internal development and external relations. At home, Khatami, buoyed by a massive electoral mandate, is presiding over a society in the process of rapid change. Intellectuals, students, women and progressive clerics are boldly questioning the political and ideological certainties of the twenty-year-old Islamic Revolution, and demanding and exercising greater social, cultural and political freedom. Abroad, Iran's rehabilitation in the eyes of the international community continues to progress. In the region, to the south there has been a breakthrough in relations with Saudi Arabia, consolidation of ties with Oman and Kuwait and at least some signs of progress in negotiations with the United Arab Emirates, Bahrain, Egypt and Jordan, while to the north links with the Caucasian and Central Asian states are growing stronger and recognition is spreading of Iran's potential as a transit country for the Caspian region's energy exports.

Relations with the West also show signs of improvement: there is a new opening with the United States and Europe's policy of 'critical dialogue' is likely to be replaced by something more positive. As of mid-1998, however, the process of rehabilitation is far from complete. Washington's Iran policy may be under increasing pressure to change, but the policy of containment and sanctions remains in force; to the south, while more voices are questioning the ability of the dual containment strategy and the US-led GCC security architecture to provide 'stable security', it has yet to be replaced and Iran is still excluded from the international security system for its own region.

This chapter analyses Iran's regional policies in the light of domestic political developments and argues that the relative success of Tehran's relations with its northern neighbours is informing its current policies towards the GCC states and may shed light on their possible policy responses.

Politics

The election of President Khatami is perhaps the most significant political development in Iran since the 1979 Islamic Revolution. In his election campaign Khatami continually stressed the importance of upholding the constitution and the rule of law, both to guarantee the people's fundamental rights and freedoms and to create a level playing field with no individuals or groups outside or above the law. He strongly advocated dialogue, openness, tolerance and pluralism as essential for healthy political and cultural development. He argued for greater social justice and for higher levels of participation and empowerment for all Iran's citizens, especially women and the young, whose aspirations have been frustrated by the regime's social and cultural conservatism and economic failures. He also argued that national interest required that social and economic issues be given the highest priority, and that the government needed to possess the requisite expertise at the highest levels to tackle these issues successfully.

None of these ideas was new to Iran's political debate, but it was the emphasis laid on them, the recognition that there were major problems in the status quo of Iranian government and society and that significant changes were called for, and the unabashed populism of Khatami's campaign that seem to have convinced voters that his was a voice that was really addressing their problems and needs and that this was a man who was committed to change. As much as the message, the way Khatami delivered it was important. Taking his campaign to the people at grassroots level across the country, he displayed a style that was radically different from the norm of Iranian politics. Spurning hollow and time-worn revolutionary slogans and pompous moralizing, he spoke simply about the issues that affected ordinary people.

In the May 1997 elections 90 per cent of the Iranian electorate turned out to vote, and almost 70 per cent of them backed Khatami. His victory came as a surprise to analysts both within and outside the country, who had seen him as the outsider in a race they expected to be won by the conservative Speaker of Parliament, Ali Akbar Nateq Nuri.

In the year since his election President Khatami has continued to surprise. Following his inauguration in August, he passed his first major test by winning the endorsement of parliament for the whole of his new team of ministers, including several outspoken reformers. The appointment of two of the latter to the Ministries of the Interior and of Culture and Islamic Guidance indicated where the main thrust of reform would go. The first months of the new government witnessed a marked relaxation in control over political expression and activity. Though Khatami himself is mild-mannered, advocates a smooth and gradual reform process and seeks to avert conflict with his political opponents within the regime, the high expectations and activism of his supporters – especially the young – have given a powerful impetus to reform. This radicalism has found expression in public political debate, which has tackled subjects taboo only a year before, and in street demonstrations, in which Khatami's supporters have shown a growing willingness to confront the *hizbullahis* and other opponents.

These developments have deepened the divisions between different factions within the regime and brought them out into the open. While the highest level of leadership – President Khatami, former president and current head of the Expediency Council Ali Akbar Hashemi Rafsanjani, and Leader of the Revolution Ali Khamenei – have succeeded in maintaining a facade of unity and have defused several explosive confrontations between progressives and conservatives, the conflict between these factions has intensified. In the summer of 1998, battles were being fought over the future of Tehran's progressive mayor, on trial for corruption and abuse of power, and over control of the Interior Ministry – the progressive minister was impeached by parliament, only for Khatami to appoint him vice-president instead. The autumn 1998 elections to the Council of Experts (a body with considerable constitutional powers) and to local authorities will provide a further trial of strength and popularity for the factions, and a further test for the cohesion of the regime.

Foreign policy

If the most dramatic and significant developments of Khatami's presidency have been in the domestic political and social arenas, the new government has also adopted bold foreign policy initiatives. The new Foreign Minister, Kamal Kharrazi, holds a US Ph.D. and for a number of years represented Iran at the United Nations in New York – a particularly important post, given the lack of diplomatic relations between Iran and the United States. His first statements, especially his September address to the 52nd Session of the United Nations General Assembly, gave early indications as to the priorities for Iran's foreign policy under the new government. Again, a number of key and interrelated themes emerge, which, if not entirely new, have achieved a new prominence.

Kharrazi has described independence and national interest as underlying Iran's foreign policy. These central priorities inform and link the other goals and detailed policies. Multipolarism, meaning relations with all states (except Israel) that recognize and respect Iran's independence, while resisting hegemonism and domination, is an important aspect which underpins the emphasis on cooperative relations with Islamic states, with regional states and with Asian, African and European states. At a more detailed level, Kharrazi spoke of promoting mutual confidence-building and security through regional cooperation, of excluding the presence and influence of foreign powers, of reducing tension at regional and international levels through active participation in regional conflict resolution, peacemaking and peacekeeping.

His emphasis on economic relations – promoting non-oil exports, using diplomacy to further Iran's economic and commercial interests, exploring possibilities for international cooperation in hydrocarbon development – harmonizes with the new government's emphasis on economic priorities at home.

Several of the concepts stressed at the UN General Assembly address are clear foreign policy reflections of Khatami's domestic priorities: institutionalizing international law as the foundation of global order and the final arbiter among states, and promoting dialogue among civilizations. Iran's attachment to radical goals has not entirely disappeared, and Kharrazi reaffirmed Tehran's continuing commitment to the oppressed, especially the Palestinian cause, but on balance it is clear that the emerging vision of Iran's global role and foreign priorities places markedly less emphasis on anti-Americanism and Islamic radicalism.

The early indications provided by the Foreign Minister were reinforced by subsequent developments. One key moment was the December 1997 Tehran conference of the Islamic Conference Organization (OIC), when President Khatami's speech developed these themes, and when the extent of the Islamic Republic of Iran's acceptance by other Islamic states was strikingly displayed in the high level of attendance and the universally positive evaluation of the meeting. Its success was the more marked, following immediately after the ill-attended and soon forgotten US-sponsored conference in Qatar the previous month. Another important event was President Khatami's unprecedented January 1998 interview with CNN, which marked a radical break with the adversarial tradition in Iran–US relations. Khatami paid tribute to the achievements of the American people and appealed for dialogue between Islamic and Western civilizations and for contact and exchange between American and Iranian societies. If the CNN interview did not immediately transform Iran–US relations, it undoubtedly broke the rhetorical ice, with both President Bill Clinton and Secretary of State Madeleine Albright eschewing antagonism in their subsequent speeches and statements on Iran. It also had a strong impact on the US debate over Iran policy and contributed to the resumption of non-governmental contacts, with a round of sporting and academic exchanges in 1998.

In statements, speeches and interviews on foreign policy both Khatami and Kharrazi have laid a heavy stress on regional relations. In his address to the UN General Assembly Kamal Kharrazi stated that: '[the] highest foreign policy priority ... is to strengthen trust and confidence and peace in our immediate neighbourhood'.

The rest of this chapter examines this priority in terms of its significance for the GCC states, with reference to Iran's regional relations to the north, with the states of Central Asia and the Caucasus.

Regionalism in Iranian foreign policy

The emphasis on regional relations is by no means a new phenomenon. Within Iran's foreign policy establishment over recent years it has been gaining prominence in theoretical discussion of foreign policy goals and methods, as well as in policy statements. The conceptual framework for the developing theme of regionalism comprises a number of interrelated ideas.

Regionalism is seen as a necessary concomitant of globalization, which is itself viewed as a process of increasing human interdependence, taking place in the framework of a set of interlinked and overlapping regions (geographic, economic, civilizational). The emergence and reinforcement of these regions and their internal linkages are, therefore, a part of the globalization process. Among those regions, mutually beneficial economic relations, a dialogue of civilizations, and tolerance and moderation are viewed as the basis for a developing global civil society. In this regard globalization and regionalism offer a way out from the historical pattern of confrontational and unequal relations among great powers, and between them and the smaller states that have been the target of their imperialist or hegemonic ambitions.

Regional cooperation, both bilateral and multilateral, also picks up on an old theme of the left wing of the Islamic Revolution, which laid a strong emphasis on relations with developing countries as an important foreign policy strategy. The argument is that such cooperation helps to guarantee independence, in the sense of both sovereignty (helping less powerful states to resist external domination through solidarity) and self-sufficiency (mutually beneficial cooperative economic relations helping to reduce their dependence on trade and investment with the great powers).

A key concept has been that regional security can best be guaranteed through regional cooperation and collective participation. The states of any region, it is argued, have the strongest stake in its stability and security, since it is they who are the first to suffer when conflict breaks out. By the same token, it is argued that regional organizations and fora offer optimal vehicles for the resolution of conflicts and disputes through negotiation. The same arguments provide justification for Iran's demand that external powers should be excluded from regional security arrangements, since they pursue their own interests rather than those of the region and tend to exert a destabilizing influence.

In addition to the theoretical justifications for prioritizing regional relations, it is obvious that the regionalism theme also serves Iran's interests in more pragmatic ways. Iran's geographical position, size and economic and military muscle give it the potential to play a leading or pivotal role in a number of regional configurations – the Gulf, Central Asia, the Islamic Conference Organization, and the Economic Cooperation Organization, among others. Since the collapse of the Soviet Union there has been a growing awareness in Iran of the country's possible opportunities as a transport and communications hub at the heart of the Eurasian continent.

The development of regional relations has also been a key element in Iran's efforts to frustrate the US policy of containment. Ideally, Iran would like to see the exclusion of outside powers, especially the United States, from a major presence or role in its neighbourhood, but the more modest and achievable goal is to make Iran a player in all regional structures and organizations.

Finally, the promotion of regionalism, and of the kind of cooperative mutually beneficial state-to-state relations it requires, has played an important part within Iran's foreign policy establishment in helping to elaborate new priorities, new revolutionary principles more appropriate for the contemporary situation and the opportunities offered by Iran's geographic location. In this respect, it has played a part in reconciling some of the old contradictions in the Islamic Republic's foreign policy – the ideology versus national interest dichotomy, for example.

Iran and the Gulf Cooperation Council

The new government has explicitly stated on several occasions that top priority will be given to relations with the GCC states. Speaking at the 52nd UN General Assembly, Foreign Minister Kharrazi expressed the desire to 'turn a new page in relations' and 'undertake new initiatives in a more energetic fashion'. He returned to a familiar theme by arguing that money spent by Gulf countries on arms could be better spent on economic development, which would also create confidence in the region. He held out an olive branch, saying: 'Let's be transparent. Let's sign a pact of non-intervention and non-aggression. Let's cooperate for economic development of the region.' He also emphasized the urgent need for measures to defuse tension. Placing relations with the GCC firmly in the framework of Iran's conception of regionalism, he stressed that: 'Our [regional] policy is based on good-neighbourliness and promotion and consolidation of mutual trust through the adoption and pursuit of confidence-building measures.'

Kharrazi's words in New York have been reinforced by other statements. For example, Deputy Foreign Minister Muhammad Javad Zarif stated that Iran was ready to start 'a process of establishing a mechanism of security cooperation which should include all the countries of the region and be based on the principles of non-intervention, mutual respect for one another's sovereignty and mutual consultation.'[1]

The priority accorded to relations with the GCC states has been reflected in exchanges of visits: thus Kamal Kharrazi's first major trip as foreign minister was to the GCC states in November 1997. With Bahrain and most notably Saudi Arabia, Iran's efforts to overcome the suspicions and difficulties of the past and 'turn a new page' have begun to bear fruit, though these initiatives, especially with regard to Saudi Arabia, predate the new government by some time. The Iran–Saudi rapprochement has been marked by high-level visits, and by a radical change in the tenor of relations. In the past the two states were bitter rivals for leadership in the Islamic world, and used fora such as OIC or OPEC to trade accusations. Ayatollah Khomeini had stated that while he might eventually come to forgive Saddam Hussein, he would never excuse King Fahd his perceived transgressions. Since 1997, however, the past appears to have been forgotten, as a variety of motives (from anxieties over falling oil prices to the need to respond to public dissatisfaction with US and Western policies towards Iraq and the Israel–Palestine question) have impelled both Tehran and Riyadh to develop cooperative rather than adversarial relations. Whether at OPEC or OIC, a clear willingness to share a regional leadership role has emerged.

With regard to the other GCC states, while Iran's initiative has on the whole been welcomed, there are still obstacles to a significant improvement in relations, and there is no early prospect that the GCC states will begin to question the current regional security structure (based on US-led containment of the Iranian and Iraqi threats) and seek to replace it with the kind of exclusively regional security architecture that Iran has been proposing.

Looking back over the past several years, it is possible to detect a

[1] *Ettela's International*, No. 842, 17 October 1997.

pattern of gradual improvement in relations between Iran and the GCC states. The pattern is, however, far from uniform, with marked variations in the character of relations with the various GCC states. Moreover, progress has been fitful and slow, inadequate to defuse the high degree of tension and mistrust that have characterized relations ever since the Islamic Revolution.

Several GCC states have serious concerns about relations with Iran, and about the possible dangers involved in a positive response to Iranian overtures. In part these concerns stem from the great disparity in size and power between Iran and the small Arab states on the opposite shore of the Gulf. Another factor is the long history of Arab–Iranian rivalry and competing claims over the Gulf – a rivalry which is played out even over the name of the stretch of sea whose shores they share. The dispute between Iran and the United Arab Emirates over Abu Musa and the Tunbs islands remains a major source of tension in relations, one which diplomatic efforts in 1998 may have helped to defuse and regulate, but which remains resistant to resolution. Past and current contradictions in Iranian policy have also played their part, with many GCC leaders believing that Iran's deeds (the escalation of the islands dispute; the military build-up and naval exercises; the record of support for Islamic radicals in GCC states) speak louder than the words of peace and cooperation.

Some of these concerns have nothing to do with Iran's Islamic regime – the disparity in size and power, the long history of Arab–Iranian rivalry and the territorial disputes, for instance, would affect relations irrespective of the ideological nature of the regime. But the ideological differences have made a big contribution to the tension – not only directly (through Iranian support for Islamic radicals, for example), but also indirectly, in that it is much harder for political elites from such different ideological traditions to establish relations of trust and goodwill. In spite of President Khatami's good public image and the striking growth of trust between the Iranian and Saudi leaders, for the GCC states the question 'Can Tehran be trusted?' still lies at the heart of the central dilemma of how to manage relations with a much more powerful neighbour, with which there is a long history of rivalry and competing claims. Iran may denounce the hegemonic behaviour of external powers, but to the small GCC states, Iran looks ready at the first opportunity to slip into the role of regional hegemon.

The question 'Can Tehran be trusted?' can be broken down into a further set of questions, none of them easy to answer.

- Who controls policy? The Islamic Republic possesses a unique constitution, and its politics and decision-making function in ways that remain opaque to outsiders, though under Khatami they are becoming more transparent. As regards foreign policy, it is hard to pin down the locus of decision-making or identify which of the numerous actors in the foreign policy process has the decisive voice, or what are the dynamics between them. Even if Khatami's and Kharrazi's statements are accepted at face value, what guarantee is there that they will be the only or even the driving force in policy formation and implementation?
- Is policy consistent or coordinated? Ideological and factional differences within the regime and the international engagement of a range of regime and regime-affiliated actors have hampered the pursuit of clear foreign policy objectives. Often Tehran has appeared to be simultaneously pursuing contradictory policies, or to be saying one thing while doing another. Is there any likelihood that this will change under the Khatami government?
- Are policy and behaviour responsive to either engagement or containment? This question is also the central issue in Western debate over Iran policy, with the United States and Europe adopting different positions, as discussed in previous chapters.

In the following pages, it will be argued that at least partial answers to these questions can be found in an analysis of Iran's relations with the Central Asian and Caucasian states across its northern border. Here too a number of relatively small states have been obliged to develop policies towards the Islamic Republic of Iran, while the latter's policies towards them have been formulated and implemented in the context of its declared priority of cooperative regional relations.

Iran and its northern region – Central Asia, the Caspian and the Caucasus

It is still not customary to think of the countries of the former Soviet South in the same context as those of the Gulf, so it may be helpful to explain how Iran's regional policy to the north has relevance for its regional policy to the south.[2] First, the development of relations with the Central Asian and Caucasian countries was, in effect, the workshop in which the current model for Iran's regional policy was created. That model, in turn, was shaped by the failure of Iran's policies in the Gulf in the 1980s and early 1990s. Conversely, experiences with the northern neighbours have come to inform directly Tehran's efforts to engage the GCC.

Second, the Central Asian and Caucasian states are, like most of the GCC states, small and weak relative to Iran. For partly similar reasons, many of them have chosen to remain reliant on a big power (Russia) for their military security. Russian border guards still patrol the external borders of most of them, Russian military bases are located on their territory, and they are engaged with Russia through a series of bilateral and multilateral treaties and security agreements. It is worth bearing in mind that the Soviet Union, in everything but name, pursued a firm policy of containment of Iran, making its southern republics a no-go area for Iranian interests and involvement.

Further similarities between Central Asia and the Caucasus on the one hand and the GCC on the other may be found in the elite and popular suspicions of Iran and its ideological and regional ambitions. All the Central Asian and Caucasian states have adopted secular constitutions and most are led by members of the old Soviet elite, who reject outright any possibility of allowing religion a role in politics. At the same time, while radical Islamic opposition movements in the Central Asian and Caucasian states are for the most part still in their infancy, they do exist and are suspected of seeking external support. Only in Tajikistan has the Islamic opposition emerged as a major force (and Iran did lend it support,

[2] The linkage between Iran's Central Asia and GCC policies was noted as long ago as 1992: see R.K. Ramazani, 'Iran's Foreign Policy: Both North and South', *Middle East Journal*, Vol. 46, No. 3, Summer 1992, pp. 393–412.

at least for a time in 1992–3), but other states, notably Uzbekistan and Azerbaijan, have cause for concern on the same score.

Furthermore, the Caspian region, containing the world's largest unexploited oil and gas reserves, is currently the centre of intense international interest. US, European, Russian, Japanese and Chinese commercial and strategic interests are increasingly engaged there, lending the region some of the geopolitical and geo-economic characteristics of the Gulf. Iran has the potential to act as a transit country for the region's economic links with the global economy. In particular it is competing with Russia, Turkey, Afghanistan and China to secure backing for pipeline projects across its territory. This presents dilemmas not only for the regional states (which have to weigh the economic, security and political pros and cons of the various possible export routes), but also for Western governments, in particular for the United States, whose policy has remained one of opposing the use of Iranian export routes for Caspian energy, though the economics and stability of the Iranian route clearly attract US commercial interests in the region.[3]

While it would be a mistake to exaggerate the strength and depth of these similarities, they give rise to interesting perspectives between Iran's northern and southern regional relations.

Iranian policy

Formation

The late 1980s and early 1990s presented Iran with dramatic developments affecting its global and regional position and foreign policy potential. The end of the Iran–Iraq war, the death of Ayatollah Khomeini, the break-up of the Soviet Union (leaving the United States as the world's only superpower), and the second (1990–91) Gulf war presented a series of challenges and opportunities. At a time when the regime was confronting questions of leadership and overall direction, it faced an international situation in which two of its traditionally most threatening though familiar

[3] T.R. Stauffer, *'The Iranian Connection': the Geo-Economics of Exporting Central Asian Energy via Iran*, Boulder, Colo., International Research Centre for Energy and Economic Development, Occasional Papers, No. 29, 1997.

neighbours (Iraq and the USSR) had been neutralized, at least temporarily, while a whole set of new and unfamiliar problems loomed with Iraq's uncertain future, the new configuration in the Gulf, the US government's dual containment policy, the continuing civil war in Afghanistan, and the highly unstable and unpredictable situation in the new states to the north in the Caucasus and Central Asia – regions about which Iranian government circles knew little. At the same time, mounting economic difficulties severely curtailed the resources that Iran was able to channel into foreign initiatives. These developments elicited new and more modest foreign policy priorities; there was an emphasis on establishing or re-establishing positive relations with the configurations of regional states to north and south, and on frustrating US efforts to isolate Iran internationally.

Early policy towards the independent Central Asian states was hesitant, cautious and reactive – perhaps best exemplified in the Tajikistan policy of 1992–3, when Iran first backed, then withdrew effective support from, the Islamic opposition. Fortunately for the Iranian foreign policy establishment, the more radical actors in the Islamic Republic's international relations proved no better informed and relatively little interested in the region, leaving the field open to President Rafsanjani and the Foreign Ministry to identify interests, formulate priorities and develop policies.[4] These can be summarized as follows.

Security

Conflicts and instability in the Central Asian and Caucasian countries had the potential to threaten Iran's security and even national unity and territorial integrity through a direct spillover of fighting, refugees or collapse in one of the new states; or a crisis in ethnic or sectarian relations (especially if involving Tajiks, Azeris or Turkmen). Tehran soon came to the conclusion that these dangers significantly outweighed any potential benefits in supporting ideological or ethnic clients.

Iran's policy response has been to try to promote stability and economic cooperation, creating common interests and thereby lessening the potential for tension and conflict. It has sought to play a constructive role

[4] Edmund Herzig, *Iran and the Former Soviet South*, London, RIIA, Former Soviet South Key Paper, 1995.

in conflict resolution – in the case of Karabagh its initiatives have been marginalized by the OSCE peace process, but Tehran has made a major contribution to the peace process in Tajikistan. It has consistently emphasized the principles of territorial integrity and non-intervention in other states' internal affairs and has preached the virtues of tolerance and cooperation, against the vices of extremist and separatist nationalism.

Security concerns have persuaded Iran to acquiesce in a strong and continued role for Russia, at least in military/security matters, as the best guarantor of regional stability and the only power capable of negotiating lasting ceasefires (as demonstrated in Karabagh and Tajikistan).

Geopolitics

The possible influence or domination over the Central Asian and Caucasian countries of a hostile major power has been a serious concern for Iran. A resurgent Russia could pose such a threat, but viewed from Tehran that currently seems a remote risk. Iran has, therefore, expressed no anxieties about the numerous CIS military/security agreements or the bilateral agreements in this field between Russia and the Central Asian states.

Far more threatening is the possibility of the United States gaining a powerful voice in Central Asian and Caucasian capitals, either directly or through what Iran sees as its proxies – Turkey and Israel. It was the US-supported Turkish foreign policy initiative in Central Asia in the early 1990s that first galvanized Iran's Central Asian diplomacy. Tehran, both officially and through the media, has expressed its anxiety at each new sign of Western involvement or interest in the region. This applies to investments by Western, especially US, companies; to the Central Asian countries' developing economic and political relations with the West (especially with the United States and Israel); and most strongly to any security dimension to those relations – recent examples are Tehran's responses to the possibility of US support for Kazakhstan's Caspian flotilla (which was denounced as a plan to establish a US naval presence on the Caspian), and the media comment on the September 1997 US-led military exercises in Central Asia within the NATO Partnership for Peace framework. Iranian fears, however unrealistic they may be, appear to focus on the possible replay of the Gulf scenario, with the United States developing strategic interests in the region and coopting the small regional states into a US-sponsored security structure directed against Iran.

Iran's response has been to attempt to engage the Central Asian and Caucasian countries, and to persuade them of Iran's good-neighbourly intentions and reliability as a partner; and again, to rely on Russia to thwart the intervention of other powers in its near abroad. As a senior research fellow at the Iranian Institute for Political and International Studies expressed it: 'Iran also expects Russia to stand by its commitments and, as a powerful neighbour, prevent the influence, intervention and ambition of foreign powers in the region'.[5]

Economic

Iran has consistently emphasized the economic dimension of its relations with all the Central Asian countries. There are a number of factors underpinning this emphasis:

(1) Iran's own economic problems and the need to develop non-oil exports have made the search for new markets and trade partners an important priority, with the northern neighbours appearing to offer some potential in this regard.

(2) Trade and economic relations offer a non-threatening way for Iran to engage with the Central Asian and Caucasian states and persuade them that their Soviet-era or US-inspired suspicions of Iranian intentions are ill-founded.

(3) Iran is favourably located to play an important transit role for the land-locked Central Asian and Caucasian countries, offering access to the Gulf via its roads, railways and ports, and potentially to world markets for their energy exports, by a combination of swap arrangements, construction of new infrastructure, utilization of Iran's existing infrastructure, creation of free trade zones, and so on. Iran sees itself as constituting the vital link between the energy rich Caspian and Gulf regions, a position it can exploit to the benefit of the energy-producing countries of both, while Iran

[5] M. Mohsenin, 'Iran's Relations with Central Asia and the Caucasus', *Iranian Journal of International Affairs*, Vol. 7, No. 4 (Winter 1996), p. 851.

itself stands to benefit commercially and to develop interdependent relations that will make containment impossible.

(4) Mutually beneficial cooperation with Central Asian and Caucasian countries in the fields of oil and gas exploration, production, refining and export is proposed on the grounds that Iran has something worthwhile to offer in the way of experience and expertise.

Iran has pursued these objectives by negotiating a large number of economic agreements, by opening new border crossings and establishing free trade zones, and by infrastructural development, notably the road, rail, electricity grid and pipeline links with Armenia and Turkmenistan. Plans for further rail development inside Iran (the Mashhad to Bafq line, for example, will cut 900 km from the route from Central Asia to the Gulf) are also directed mainly to improving Iran's potential as a transit route for Central Asia. A large number of joint ventures have been established between Iranian and Central Asian and Caucasian firms, and trade is growing vigorously (though from a very low starting level).

As an economic partner Iran is severely constrained by its economic, especially manufacturing, weakness and by difficulties in raising international finance for projects of all kinds. It enjoys the advantage of geographical proximity, but except as a transit country Iran has, and is likely to continue to have, difficulty in competing with its main commercial rivals – Turkey, Russia, China and the more advanced Western and Asian economies. This relative weakness, however, does not mean that economic engagement in Central Asia is insignificant for Iran. Even quite small-scale non-oil exports appear significant in the context of this poorly developed sector of Iran's economy. In the long term Iran is likely to play an important role in the region's developing transport and communications networks, though in the short term the realization of that potential will be hampered by US opposition.

Cultural

The historical and cultural closeness of Iran and Central Asia has been a recurring theme in Iran's Central Asian diplomacy. The emphasis has been carefully placed on literary, scientific, artistic and other ties, i.e. on the classical Islamic culture, expressed largely through the medium of the Persian language, that was spread through much of the region in the period from about the eleventh to the eighteenth century CE – rather than on Islam itself, let alone revolutionary Islam. Even so, this diplomacy has frequently backfired, as Iran's implicit or explicit assumption of leadership and its offer of a guiding hand to societies supposedly needing to rediscover their forgotten cultural roots have raised hackles among Central Asian political and intellectual elites, who have their own interpretations of their histories and cultures.

Apart from emphasizing culture in official dialogue, Iran has tried to promote the Persian language, has sent books and periodicals, and has tried to encourage academic and cultural exchanges and joint activities. This has been a particularly strong element in the case of Tajikistan, the only persophone country in the region.

Regional cooperation

Regional cooperation has been a central concept in Iran's policy towards Central Asia and the Caucasus. Officials and analysts continually emphasize the benefits of regional cooperation, both bilateral and multilateral, in promoting trust, security, stability and economic growth, in safeguarding common interests and in preventing external domination. There is an undisguised political agenda to this emphasis. By integrating as fully as possible into the Central Asian–Caspian–Caucasian region Iran hopes to make isolation or encirclement by the United States difficult or impossible.

While regional cooperation is emphasized in bilateral relations, perhaps the most significant development has been Iran's efforts to launch new or rejuvenate old multilateral organizations for regional cooperation. Two of these in particular are worth more detailed consideration:

(1) The 1992 expansion of the Economic Cooperation Organization (ECO) to include the Central Asian states, Azerbaijan and Afghanistan was primarily an Iranian initiative. If the more grandiose ambitions for this 'Islamic Common Market' appear premature or over-optimistic, and if it has served sometimes to reveal rather

than to overcome the differences between its members (President Karimov of Uzbekistan, for example, threatened to quit the organization if certain countries, by implication Iran, continued to exploit it as an ideological platform), nevertheless, the ECO initiative has helped to consolidate Iran's place among its neighbours to the north, northwest and east (in contrast to its continuing exclusion from the Gulf 'club' to the south). ECO at the very least has the potential to play a practical coordinating role in such areas as prioritizing and planning international transport and communication schemes.

(2) Iran's other major initiative was the 1992 proposal for a Caspian Sea Cooperation Organization (CASCO). While the organization has not yet fully come into being – largely because of the deep differences separating the Caspian countries on the most important issues of legal status and exploitation of sub-sea energy resources – the initiative was timely, in underlining the necessity of cooperation and agreement among the Caspian countries on many issues relating to the sea and its resources. It has been followed by a number of meetings of officials up to foreign minister level from all the littoral states. The 1992 CASCO initiative also gave Iran the opportunity to place on the negotiating table its 'condominium' solution for the Caspian Sea's legal status and energy resources, as well as its proposal for a demilitarized Caspian. The condominium solution was closely identified with the Russian Ministry of Foreign Affairs from the advent of the September 1994 Azerbaijan International Operating Company (AIOC) consortium contract until 1998.

In most analyses of the Caspian issue, Tehran is presented as being supportive of Moscow, and for reasons to do with overlapping objectives – such as a desire to maximize its share in the sea's energy resources and to have the power to veto unwelcome foreign penetration into the region – but the Iranian proposals predated Russia's expression of them and are better viewed as an independent statement of Iran's own priorities for the sea. Financial benefits from the production and export of oil and gas are one concern, but hardly the central one. The CASCO proposal was intended rather to guarantee Iran a role as an equal player in the Caspian game, and to create a regional organization capable of withstanding external (i.e. US) pressure and intervention. Iran's priorities on the Caspian Sea issue are demonstrated by the fact that it has continued to insist that the legal status issue can only be resolved by unanimous consent of all the littoral states and agreement on the illegality of unilateral action, whereas its preference for a 'condominium' solution has not always been so strongly stated (it seemed almost forgotten during the period when it seemed likely that Iran was going to secure a share in the AIOC contract, and is under review and will surely shortly be discarded now that Moscow has altered its position).

Iran's Caspian policy is actually informed by its experiences in the Gulf, where the involvement of Western oil companies led to the identification of the Gulf as an area of vital Western/US national interest, with all that followed from that. Seen in this light, Iran's enthusiasm for CASCO, for a condominium status and for demilitarization can be seen to have a political and security rationale rather than an economic one. This also explains both Iran's readiness to join Caspian consortia, although they contravene its interpretation of the legal status issue (the most important priority is to be a player, and not to be excluded and isolated), and the intensity of its frustration when in spring 1995 Baku bowed to US pressure to exclude Iran from the AIOC contract (which seemed to presage the beginning of a Gulf scenario for the Caspian).

Summing up

Iran's foreign policy towards the Central Asian and Caucasian states has undergone significant development in the years since the collapse of the Soviet Union. The period has seen the emergence of clear priorities of cooperative state-to-state relations; the pursuit of regional security and stability; the development of mutually beneficial economic relations; the promotion of regional organizations and frameworks for engaging with the Central Asian and Caucasian countries. In its strategy to the north, Tehran has opted for and pursued a policy of positive engagement in order to secure its national interests and frustrate the US policy of containment. Tehran has not always been able to achieve its goals and has had to moderate early ambitions, while frictions remain in relations with certain states (notably Azerbaijan and Uzbekistan), but on balance the strategy has been paying off. Iran has gradually become more fully engaged with its northern neighbours; suspicions of Iran's intentions have tended to diminish; and there has been increasing recognition of Iran's capacity to contribute to the region's economic development and play a positive role in conflict resolution and the pursuit of stability.

A number of factors have contributed to the very different trajectories of Iran's relations with the Central Asian and Caucasian countries on the one hand, and with the GCC states on the other. One factor is purely historical: because the Soviet Union did not break up until 1991, Iran's relations with the new states to the north grew up in a period when its revolutionary zeal had already been tempered by more than a decade of experience, and at a moment when the need to respond to the US dual containment policy made Tehran very careful to seek out and hold on to friends.

Perhaps the single most important factor has been the role played by Russia. The major regional power in the Central Asian and Caucasian region, Russia, for a complex set of reasons, deliberately adopted a policy of positive engagement with Iran. Tehran was only too ready to reciprocate, and maintaining good relations with Moscow became a key priority, giving a major incentive not to annoy Moscow by causing trouble on its southern flank – even to the extent of acquiescing in continued Russian regional hegemony. Influenced by Russia's stance, the leaders of Central Asian and Caucasian countries have gradually come round, persuaded in part by Iran's consistent non-threatening behaviour, and in part by the value of Iran as a transit route to the south or west.

In the Gulf, by contrast, the United States remains the dominant power and has, until very recently, consistently discouraged the region's states from engaging with Iran, promoting instead the status quo and the GCC security structure, designed around the policy of dual containment. Iran, meanwhile, has continued to call for the exclusion of external powers and the establishment of a security architecture involving the regional states alone.

There are obvious and important differences between the regions to the north and south of the Islamic Republic, but the example of Iran's relations with Russia, Central Asia and the Caucasus demonstrates that in the right circumstances Tehran is willing to reciprocate in cooperative, non-conflictual relations, and that the existence of tensions and suspicion need not lead to a downward spiral in relations. In its regional relations to the north, Iran has shown that it is responsive to engagement, and that it is capable of pursuing a more or less coordinated and consistent foreign policy.

8 The new generation in the GCC: the case of Saudi Arabia

Mai Yamani

In all the member states of the Gulf Cooperation Council population growth rates are high. The 'baby boom' which occurred in the wake of the 1970s oil boom has produced a transformation in the demography, with over 40 per cent of the population under 15 years old and only 4 per cent over 60. Not enough is known about the values, motivations and aspirations of the new generation to assess how they will respond to the challenges ahead. This is the subject of discussion here and the focus is on the situation in Saudi Arabia in particular, as the largest oil producer and leading regional power in the GCC.

The central feature of the political systems in the GCC states is the traditional royal family who, but a generation ago, were desert tribesmen. Across the GCC these ruling families still rely, to varying degrees, upon Islamic and tribal values to maintain their political legitimacy. Yet in each of these states the generation that has come of age in the last decade has been removed from the pastoralist and nomadic past, knowing only a situation in which income from oil has provided for urban and modern lifestyles. The generation that will come of age over the next twenty years is, however, experiencing new strains and influences, as awareness grows of the vulnerability of these societies in the face of decreasing oil revenues.

Economic imperatives and social norms

The Saudi standard of living has slipped from one of the world's highest in 1981, when per capita GNP was $12,600, to the level of a middle-income nation by IMF standards in 1993, with a per capita GNP of $6,750.[1] In 1996 the per capita income was $6,390 and the downward trend continues.[2] This has meant that governments can no longer afford lavish expenditure upon social services such as welfare, health and education. Welfare and development have been central tenets of the ruling ideology and have cushioned the indigenous populations from the sort of social and political problems endemic in the other states of the Middle East and North Africa. The IMF has warned of increasing indebtedness, and there have therefore been attempts to freeze state-sector employment in all the GCC states.

In Saudi Arabia this policy conflicts with Saudi-ization, a programme designed to train Saudi nationals to take up positions in the economy hitherto held by migrant workers. The difficulties of replacing expatriates with nationals are endemic in the GCC states, which have at the same time attempted to implement reforms to the private sector to attract foreign investment. The resulting policies, it should be noted, reflect the market values of American business society and its informal representatives, the World Bank and the IMF.[3] Although the Gulf states will continue to be of importance as a world centre for hydrocarbons, and indeed their role in global energy will increase in coming decades, the economic reforms point to social tensions that are as important to consider as are international energy market and regional political developments in the Middle East.[4]

Perhaps the greatest threat to the established governments is the size of the new generation itself. In the GCC states, by the year 2020 the proportion of the population under 15 is expected to reach 50 per cent. The rapid expansion of the population has magnified generational differences, presenting problems for the integration of the new generation in the economies, societies and states of this region. Typically, the societies of the GCC have been characterized as organic communities, organized according to shared tribal and religious norms. Yet these values

[1] *Financial Times*, 22 December 1993.
[2] *Independent*, 3 January 1996.
[3] The degree to which the World Bank's lending decisions are influenced by American political and strategic interests is discussed in Catherine Caufield, *Masters of Illusion, The World Bank and the Poverty of Nations*, Macmillan, 1997.
[4] For a discussion of the increased role in the petroleum and petrochemical industry worldwide, see 'A Look Ahead', *Gulf States Newsletter*, Vol. 19, No. 500, 28 November 1994.

overlie regional differences within and between countries and, of course, between generational groups. Some indicators show a tendency among the younger generation to reject Western modernity in favour of modern Islamist ideology. Other indicators suggest a preference for values emanating from outside the cultural tradition, giving rise to new expectations and frustrations. Though conflicting, these indicators demonstrate a pattern of social change and adjustment, with inherited cultural characteristics persisting. Examples of such attitudes include the feeling that Saudi nationals should avoid menial, or 'dishonourable', labour; that women should be segregated from male society; or that foreign workers should provide a labouring sub-class.

These inherited values are reflected in the norms of the dominant social organization of these countries: that is, the extended family – patrilinear, patrilocal and patriarchal. Although migration into urban centres has resulted in some fragmentation of the extended family, it nonetheless continues to function as a social net that spares the mother from having to provide material support and leave the home to work. The values of the extended family are in turn reflected in the Islamic heritage, which the modern GCC states actively nurture. Many Western analysts assume this policy is anachronistic and financially untenable, believing that, for instance, by accentuating Islamic values in education, especially the segregation of the sexes, the Saudi Arabian government has not prepared the population for the necessary structural adjustments that the economy and society will have to undergo in the coming decades. This analysis is present both in World Bank reports and in the journals and newspapers that rely on them. However, it is difficult to see how these Western assumptions will assist in the economic and social restructuring of GCC states in the future. Saudi Arabia is the foremost example of a GCC state that has undergone major societal transformation as a result of attempts to develop the human resources of the country, particularly by the expansion of education. It is certainly not a coincidence that Saudi Arabia has also recently witnessed demands for political and social reform.

Tradition and the state in Saudi Arabia

There are a number of reasons for turning the focus onto Saudi Arabia in particular. Its economic and strategic role in the area is crucial, but its relative religious conservatism compared with other GCC states also warrants special interest. In addition, quite simply, Saudi Arabia happens to lie at the heart of the academic work of this author.

The ruling family in Saudi Arabia, Al-Saud, has maintained a high degree of organic solidarity as a patronymic group while becoming a national elite. It controls the important government positions and has strengthened its base of support by alliance with the religious establishment, the Al-Shaykh, and the regional elites. There are 6,000 princes of Al-Saud and, combined with the Al-Shaykh, there are altogether as many as 12,000 men who are related by marriage to all the other local lineages such as the Sudayris, putting the number of the ruling class at around 20,000. Together with other non-royal partners in the kingdom's elite, the total is about 100,000. So the influence of such a substantial ruling class should not be underestimated.

The rule of the Al-Saud family represents stability and piety, and these values override the absence of representative institutions. The system of monarchic rule is a form of patronage, depending upon the immense oil wealth controlled by Al-Saud. The system has recently been described in the Western press as one in which 'no representation equals no taxation'. Tax follows the pattern of religious obligation, *zakat*, an obligatory alms tax of 2.5 per cent of surplus wealth per annum. The issue of income tax remains a controversial one because *zakat* is considered the only legitimate form of tax according to Islamic law. The royal family thus follows a traditional pattern of rule: the distribution of patronage, both in the form of influence over public-sector jobs and by the provision of welfare schemes for the remainder of the population.

The state is the largest employer, with a total of 88 per cent of the workforce employed in the public sector.[5] Therefore, access to government influence, usually by family connection, is the normal method of career advancement with the result that there is a high degree of personal dependence of the individual upon the state. This dependence might be

[5] Saudi Development and Training Company Ltd, 6 July 1996.

considered as tending to support the organic structure of society, encouraging a 'personal' relationship between rulers and the individual, reflected in the accepted method of petition writing and appeals directly to the ruler rather than formal institutions of political representation. The other area of state patronage has been the provision of free healthcare and education, and subsidies for medicine, bread, electric power and petrol. Likewise the state offers interest-free loans and gifts of land.[6] There is also the novel phenomenon of men relying upon state welfare for a $5,000 gift for the marriage dower, a form of state encouragement for marriages between Saudi nationals.

The state has thus attempted to adapt the 'traditional' social and moral practices of paternalism to the new conditions of a modern state, a phenomenon often described as the patrimonial state. This is a remarkable adaptation of tradition to the modern state. Yet, when considering the attitudes of the younger generations in society, it would seem appropriate to ask whether the moral concepts that attend patronage and charity have been transmitted to the new generations, or whether such values have been compromised by the financial constraints of the contemporary era. In the present situation the patrimony of the ruling family has become constrained, excluding a larger section of the population from the immediate benefits of state patronage. Meanwhile, the IMF has urged the expansion of the private sector, but although certain market sectors have been growing, the economy overall has remained stagnant in the recent past.[7] From January 1995, the state has been obliged to raise the price of petrol, water, electricity and telephone rates; at the same time Saudi Aramco, the state oil company, has reached the end of a long period of capacity expansion. The situation begs the question: in the absence of a complete social welfare system will society retain the organic solidarity of an Islamic and patriarchal society?

The family and values

Inherited values are determined by the most important social organization in these societies, the family. The family remains patriarchal. Men

[6] *New York Times*, 25 September 1981.
[7] 'No Growth in the Kingdom', *Gulf States Newsletter*, Vol. 21, No. 533, 3 April 1996.

dominate the family, with women bearing the enormous responsibility for rearing an average of five children. The husband, by *iltizam*, obligation, must pay for all household expenses. The wife makes only one-third of food purchases and the husband makes all purchases of durable commodities, including the satellite dish! Therefore, although Saudi society has become highly commercialized, with 95 per cent of households owning a television and 60–70 per cent a stereo or video, the patriarchal character of the family should reduce the impact of consumerism.[8]

The gender distinctions are evident in the consumer sector, with special branches of banks for women, for example, and most banks offering different bank cards for men, women and the young (those under 25). These distinctions are taken for granted by the youth. The separation of genders is also apparent in the big shopping malls, where certain shops are clearly designated with signs reading: 'For women only. Men are not allowed.' In the major trading and commercial enterprises women have their own niche through which they actively engage in the private sector. It is estimated that women hold 25 per cent of property in Riyadh and 50 per cent in Jeddah, a fact reflective of the rights afforded women by Islamic law, such as autonomous control over inherited wealth and dowers. Assuming the development of the private sector, will a new generation of women seek to become more integrated in the market and what implications will this have for a society organized along patriarchal lines? Already women educated in the universities and schools provided for them are seeking positions in the workforce behind the veil.

A number of factors do seem to have dramatically changed attitudes in the new generation, especially among younger women, who have made demands for rights, usually couching these in Islamic terms. The result is a mixture of modernity and tradition. For example, it is the concept of honour – family shame – that prevents women from complete freedom of business management. Likewise, a woman can be a major shareholder in a company but yet unable to attend a board of directors meeting for fear of coming into contact with men. These restrictions are often based not on Islamic principles, but on tribal norms of honour and shame.

[8] *The Saudi British Bank, The HSBC Group: Business Profile Series*, Saudi Arabia, sixth edn, third quarter, 1994, p. 15.

Values and the workplace

Because of their awareness of their economic capacity, these young Saudi women have developed what could be viewed as an Islamist-feminist movement.[9] Beginning in the universities, this movement was nurtured by lecturers who were influenced by Egyptian ideas on women's Muslim rights. The movement spread rapidly with gatherings under charismatic leaders. In their meetings, attended by as many as 300, these women remained veiled to represent their solidarity and their commitment to Islam, so distinguishing themselves both from the tradition by which women only wear the veil in public, and from liberal protesters who have less support within an Islamic society. The worldly aims of the young feminists are not yet clearly defined. They do not demand the right to vote or to work with men. However, they seem to be in the forefront of all public professions that are open to women at women's branches of banks and the universities. Therefore, although this group clearly represents the educated new generation, the Islamic motive provides them with a moral power to challenge patriarchal norms that stand in the way of their professional achievements. The aggressive attitude is a significant distinction between the new women and those of the same generation who have accepted the traditional role of women in the family. Thus their Islamic attire belies a new and even modern attitude towards work and public life.

The patriarchal structure of society has dominated development in the private sector: thus private businesses, even some of the largest ones, are family concerns. It is estimated that 99 per cent of businesses in Saudi Arabia are family-owned.[10] The family is, moreover, unwilling to allow any other Saudi national access to business information and therefore most businesses prefer to employ foreign workers. Even those Saudis who have attended the trade or vocational schools tend to set up businesses that seek government contracts and then hire foreign companies and/or workers with the expertise to carry through the project. The result is that the huge migrant class employed by the government in the state sector of the economy is replicated in the private sector, with the migrant

[9] Mai Yamani, 'Some Observations on Women in Saudi Arabia', in Mai Yamani, ed., *Feminism and Islam, Legal and Literary Perspectives*, Reading, Ithaca Press, 1996, pp. 263–82.
[10] Saudi Development and Training Co.

class making up 41 per cent of Saudi Arabia's population. The total number of expatriates in the 1990 labour force was 2,878,000 – 60 per cent of the total labour force. The percentages are not expected to change before the year 2005. The situation has become a cause of some social tension in recent years, particularly in the cities where 78.5 per cent of the total Saudi population live, because the younger generation of Saudis must compete with foreigners when seeking employment. The government responded with programmes to evict illegal workers in 1995 and 1997, as well as making stricter rules for the issuing of visas for foreign workers.

The government's concern is occasioned by a projected annual growth rate in the Saudi population of 3.3 per cent between 1992 and 2000, with an average of 125,000 Saudis entering the job market each year (a projected 700,000 in the years between 1995 and 2000).[11] However, it should be noted that the reported number of Saudi job applicants is only 28–43,000 per year, the applicants are on average between twenty and twenty-five years of age, some 10 per cent are illiterate and 2 per cent have university degrees. The numbers indicate that although the Saudis are beginning to seek work in the private sector, Saudi men still prefer the security and status afforded by government employment. In a total labour force of 4,382,500, Saudis make up only 12 per cent of the service sector, 7 per cent of the private sector and a mere 4 per cent in industry. They do take up positions in the more desirable banking sector, however, where they make up 65–70 per cent of the work force, as well as in the privately owned petrochemicals industries, though they rely on migrant workers for unskilled labour in these cases.[12]

To reduce the perceived negative social impact of the migrant community, the Saudi government has promoted higher birth rates among nationals as well as expanding education to enable nationals to take up diverse roles in the economy. This has changed the shape of Saudi society, creating over the past generation an urban middle class of Saudis employed in the bureaucracy, numbering in the hundreds of thousands (in 1980 the number employed in the civil service was 173,290).[13] The

[11] World Bank, World Tables, 1994.
[12] World Bank, World Tables, 1994; *MEED* Special Report, Saudi Arabia, *MEED*, Vol. 40, No. 14, 5 April 1996, p. 56, and Saudi Arabia, *MEED*, Vol. 39, No. 10, 10 March 1995, p. 40.
[13] Mordechai Abir, *Saudi Arabia in the Oil Era*, London, Croom Helm, 1988, p. 170.

ruling family dominates this sector by employment contracts and control of positions in the civil service and state utilities. But the current policy of Saudi-ization represents a more active step to involve Saudis in the private sector.

Saudi-ization plans

The sixth five-year plan (1995–2000) stated its aim as to 'replace non-Saudis with appropriately qualified and willing-to-work Saudi manpower in all occupations' and to develop 'Saudi manpower through meticulous evaluation of educational curricula and training programmes'.[14] The government's policy of Saudi-ization, including training Saudi citizens for jobs in the lower levels of the economy, envisages the creation, between 1995 and 2000, of 660,000 jobs, 319,000 of them by displacing foreign workers. Since the government has frozen state-sector employment as part of a budget austerity programme, the jobs have to be created in the private sector. This may be impeded by practices in business that reflect the familial structure of society, as well as local restrictions on the movements of foreigners, altogether making investment an uncertain and unattractive endeavour for foreign investors.

Although the Saudi analysts were optimistically estimating a 4.5 per cent per annum growth in the private sector when they made these plans, one of the problems is that the younger generation of Saudis has inherited the preference for secure government positions over the difficulties and uncertainties of a career in the private sector. Reflecting this there is a preference for jobs in the oil sector and little interest on the part of the Saudis in investing in the private market sector. Without foreign investment Saudi economic growth will remain, at best, around 2 per cent.[15] Obviously, with 3–4 per cent population growth a projected annual increase of 1–2 per cent in GDP will not be adequate.

[14] Saudi Arabia, *MEED* Special Report, Vol. 39, No. 10, 10 March 1995, p. 40.
[15] Institute of International Finance, Washington DC, 8 November 1994.

Education

Although the formation of an educated and urban middle class is the most significant change to the social structure of Saudi Arabia, among this class 'traditional' attitudes persist. The new generation of men follow a pattern of patriarchal authority transmitted by the older generation, represented at the social apex by Al-Saud itself. Will this pattern be modified? Will traditional norms of respect for authority be maintained? Education is one means at hand to form the attitudes of the young and in the 1970s and 1980s the government expanded the number and size of universities, whose graduates were then absorbed into the state bureaucracy. Free education, housing, food and books were provided by the government, and until recently a job was also guaranteed. A study by the IMF of Saudi society in the 1970s reportedly showed that most Saudis with an education desired a stable, government job, because this was the most secure base from which to develop family life. Most students thus adopted a more conservative attitude to career advancement by seeking government patronage and support. The government encouraged the conservative tendency by giving the religious establishment (*'ulama*) supervisory controls over educational policy, even in the secular universities, resulting in a strong Islamic influence in university curricula. Not only did education tend to turn out people with conservative attitudes, but the teaching of foreign cultures and beliefs was restricted, and foreigners were excluded from the state schools.

The education system in Saudi Arabia, though greatly expanded, does not appear to integrate the diverse social groups of Saudi society. Rural and urban lower classes tend to drop out of the system. Primary education is mandatory, however. As of July 1996 there were 920,000 males and 769,000 females in primary education, of whom many dropped out or repeated years; thus at the intermediate level there were only 280,000 males and 217,000 females, and at the secondary level 127,000 males and 114,000 females. In universities there were 73,000 men and 57,000 women. Note that only one-third of pupils pass from primary to intermediate education, then half go on to secondary and half again to university.[16] These figures were produced by a private-sector training

[16] Saudi Development and Training Co.

institute; however, Saudis in the academic field claim that the data do not reflect the huge expansion in the educational system and that the problem is not the number of students but the quality of education. The result is that Saudi youth are competing in a highly sophisticated market with underdeveloped skills. This could cause frustration and resentment.

They believe they see uncaring employers supported by devious foreigners intent upon protecting themselves by creating barriers to entry. They believe they know a double standard when they see one. With the general use of English as a discriminator in achieving employment and a generally limited capability in terms of job search and self marketing, Saudi youth often find the employment search a degrading and frustrating experience.[17]

Their inherited deference to patriarchal authority means the young turn to the government for solutions, blaming the employers, at least in the short term, and definitely prefer government employment when available. The problem becomes acute when the government policy-makers try to impose reform on the labour force, undermining the consensus between business and the ruling family upon which the Saudis have relied.

The employers, operating in a global market, prefer not to engage in the costly job-training programmes supported by the government for nationals, and claim that anyway Saudi nationals are more difficult to integrate into the global economy because of the tribally segmented nature of society: Nejdis prefer to stay resident in the Nejd, Hijazis in the Hijaz, whereas expatriate communities are mobile and adaptable. The youth apparently resent the realities of a global market- place, expressing their dissatisfaction with the necessary language qualifications, usually English, as they are nurtured in an educational system that stresses Arabic and Islamic communal values. Likewise the cultural conflict is evident in the educational system where, in 1990, one-quarter of the university students were in the Islamic universities.[18] Mordechai Abir has noted the dilemma by citing the observation of a professor in the Saudi university system: 'English and Arabic are forces with powerful symbolic valence in Saudi Arabia; they stand for modern/traditional; secular/sacred; alien/

[17] Saudi Development and Training Co.

[18] F. Gregory Gause III, 'The Middle East: A Crucial Moment', *The Washington Quarterly*, Vol. 20, No. 1, Winter 1997, pp. 145–65.

comfortable.'[19] Proficiency in English, as the contemporary universal language, is an index of the degree of adaptation of indigenous Saudi society to the prevailing conditions of a global community.

To be sure, government educational and employment policies have been moulded to maintain an Islamic world view among the new generation. The 'world view' here means the cultural references that inevitably emerge with each new generation as social and cultural norms are learnt or discarded. Thus whereas access to global mass media would tend to widen the cultural frame of reference from town and city, family and lineage, to the global community, and lay the basis for a new world view, the Saudi national project has attempted to sustain tribal notions of Arab chivalry, family honour and Islamic morality. What is the result for the development of the economy and society? One result is a segmented social map and a widening gap in the minds of the young between myth and reality. For instance, it was noted above that family structures determine the method of social relations and of business organization. With 82 per cent of businesses employing fewer than ten people, and 12 per cent comprising large companies including the oil and gas giants, medium-sized companies make up a mere 6 per cent of businesses. The figures suggest the peculiar nature of an economy in which the state can dominate other aspects of social and cultural development. A recent analysis by a bank interested in developing management and clerical skills among Saudi youth points to certain obstacles in the formation of a modern business class. Thus the study notes that any expansion of the private sector requires the expansion of business organizations beyond the family and its customary laws by developing corporate law. However, new legislation promoting Saudi-ization runs counter to this rationale because it requires that companies with twenty or more employees increase their number of Saudi employees by 5 per cent per annum or else forfeit government loans and contracts. The result is to encourage the fracturing of businesses into smaller units and frustrate the development of a strong private sector. Saudi social policy follows a conservative line and this in turn frustrates the development of a more diverse society and economy.

Conservatism incorporates a concern with social order. It is perhaps appropriate to recall the origins of the debate on population and society in

[19] Abir, *Saudi Arabia in the Oil Era*, p. 170.

Europe, which had its origins in Thomas Malthus's 'Essay on the Principle of Population' (1798), and was in essence the expression of the aristocracy's fears of the impact that the expansion of the middle and lower classes would have upon its own ascendancy. Reports in the press that the migrant community now makes up 67 per cent of workers in the cities, and that in this sector crime, drug trafficking and forgery are on the increase, reflect such fears in Saudi Arabia.[20]

Indeed, changing population patterns pose a challenge to the inherited values of Saudi society, yet it is worth noting the adaptability of culture evident in the growing number of private charities and shelters, *arbita*, created to care for the rising numbers of urban poor. The private charities are funded by *zakat al-fitr*, the charitable donations paid at the end of Ramadan, and according to the obligation of *sadaqa*, or alms in the more general sense. Some of these organizations are run by women who have adapted their traditional role to new social realities. But as the rapidly increasing Saudi population comes on to the labour market there will be pressure to compete with foreign experts and migrant labourers and thus adapt to the prevailing cultural attitudes in the global labour market. The examples chosen here show that although traditional world views persist, these are strained by new social conditions in which class, generation, gender and global culture are as important as tribe and family. The pressure of these forces combined with the decline of oil revenues will make the questions of population and employment as vital in the future to the GCC monarchies as they are at present to the other states of the Middle East and North Africa.

[20] *Saudi Gazette*, 5 December 1995.

9 Commentary and conclusions

Rosemary Hollis

A central theme in the foregoing analyses is that the place of the Gulf region in the global setting is determined by a process of interaction between external and internal forces. Its indigenous oil resources render it important in the international energy scene. However, fears that supplies could be disrupted by conflict and instability, plus the availability of new technologies, have led to the exploitation of alternative sources of energy, in turn affecting the fortunes of Gulf producers. The end of the Cold War and successive Gulf wars have opened the way for greater involvement by the United States and its Western allies in Gulf security arrangements. Yet the internal dynamics at work in both Iraq and Iran may require those arrangements to be completely reformulated.

This broad approach implies that all the actors involved must operate within a system of constraints determined by circumstances. However, to suggest that all the players are somehow equally constrained would be misleading. Some are more equal than others. Iraq in particular has been quarantined under an international trade embargo of unprecedented proportions. Iran is thwarted in its ambitions by a security regime championed by the United States which prevents it from attaining the level of regional influence that it might otherwise enjoy. For the states of the GCC, meanwhile, the prevailing regional order is not one of their own making or necessarily of their own choice.

This perception was aired in the discussion in Abu Dhabi at the time of the seminar which informed this work. According to one commentator from the Emirates, for example, those in the region are seemingly not in a position to define their own terms of reference. They have been identified as part of a global agenda formulated principally by the United States and:

> The area itself has lost its ability to formulate a solution to its own problem. Not only that, it has lost even the ability to define the problem.[1]

This assessment reflects a key concern of some commentators about the shape and direction of the prevailing and Western dominated debate about security issues in the Gulf. In addition, citizens of the Emirates have other points to raise about some specific issues touched on in the preceding analyses. This chapter gives space to both these sets of concerns and attempts to draw some conclusions, the better to understand the prevailing situation and trends in the Gulf.

The energy scene

The UAE is thought to possess up to ten per cent of known world oil reserves. This, at least, was the estimate before interest in the resources of the Caspian region took off. An accurate assessment of the size of the Caspian reserves would clearly be useful to determine the place of the UAE in the world energy scene in the future. To be more specific, Gulf producers are interested in knowing the extent to which their position could be overshadowed by the development of Caspian oil and gas.[2] According to John Mitchell, the short answer is that there is no substitute for Middle East energy resources. Some journalists have spoken in terms of very large reserves indeed in the Caspian, but actually, as of 1998, there are no accurate assessments. Given the difficulties, both political and practical, of exploiting what is there, a conservative forecast would put the figure for Caspian oil production at around 2m bpd to 2.5m bpd by 2015 and, if everything went well thereafter and there were more significant discoveries, that figure could eventually go up to 5m bpd.

Demand for Gulf oil and gas will inevitably be affected by trends in Russian and Caspian production. If resources in both areas were not to be developed, over a five-year period there could be a shortfall of up to 2m bpd or even more, which would create a hole that the Gulf might expect to fill. However, this is not a likely scenario.

[1] Dr Jassim Abdulghani Al Khulufi.
[2] Dr Atif Ajwa.

In the medium term, inventory levels clearly have a bearing on how global energy demand and supply projections are balanced. Given that in late 1997 inventories overall were at a seventeen-year low, at around 57–58 days' forward demand, there was at that time room for concern that the margin could be too tight to cover potential contingencies.[3] In fact, according to John Mitchell again, the lower inventory levels probably are sustainable. The trend with all commodities, whether soya beans, wheat, steel or oil, is towards keeping very low inventories, in order first to improve returns on capital, and secondly to use the technology which is now available to manage logistics. Information can be transferred rapidly from the market- place to the producer. Price controls do not delay the process, as they did seventeen years ago. Meanwhile, logistical capabilities have improved and the commodity can be moved much faster on demand. The significance of inventory levels was demonstrated in 1998, but not in the form of a crisis in meeting demand. Instead, the efforts of producers to boost flagging prices in an oversupplied market by cutting back production have taken time to bite, owing to the excess volume of oil 'in the pipeline' awaiting customers.

In so far as demand for Gulf oil is expected to creep upwards in the next decade or so, it is important that the necessary capital investment will be forthcoming.[4] Significant sums will be required to reach the 35–40m bpd eventually anticipated. Gulf producer countries vary in terms of the capital cost per unit of the output increases needed. Abu Dhabi is thought to be somewhere in the middle of the range, while Iran is probably at the top end, because of the complexity of its reservoirs and the need for gas injection. As for Iraq, available estimates require confirmation. In general, Middle East oil is cheaper to develop than reserves in Europe or the United States, and should be able to attract funds. However, only in some parts of the Middle East has the private sector been given access, and political and structural issues will determine which countries will call on foreign capital and which will rely on their own savings.

According to David Heard, the capital and running costs of production in the UAE and other OPEC countries may be expected to rise as the fields mature, since more work will be required to obtain the yield.

[3] Munif Othman.

[4] Idem.

Higher costs and investment requirements will eat into producer revenues. This is important to bear in mind when forecasting revenue levels, and in part accounts for Mr Heard's caution. At present, the figures for revenue levels are in any case very misleading because they are often calculated by taking OPEC production figures and multiplying by the average price per barrel. This method obviously fails to take account of rising costs and investment needs which will have to be deducted from revenue. In this connection, commentators in the Emirates are conscious of the need to take account of all such variables expected to affect UAE oil revenues, and to plan accordingly.

The Western powers and the Gulf

Oil is usually identified as the primary US interest in the Gulf region. As its import requirement has gradually increased, the United States has sought a larger proportion of its intake from sources closer to home: US imports from Venezuela exceeded those from Saudi Arabia in mid-1998. Meanwhile, a deliberate decision was made, as a result of the oil shocks of the 1970s and early 1980s, to spread the risk and diversify US sources of supply. Nevertheless, since oil is a globally traded commodity, prices everywhere are affected by flows from the Gulf and the United States retains an interest in the stability of the region for this reason.

Viewed from inside the Gulf countries, however, this line of reasoning raises two types of concern. On the one hand, it means that the importance of the area has been determined, by others, in a global rather than regional context, thereby essentially compromising the ability of Gulf nationals to develop a regional agenda of their own making. On the other hand, it involves the United States in their affairs in a manner dictated by the pursuit of US interests. From an indigenous Arab perspective, in fact, US policy in the area is driven by three broad interests, succinctly summed up as 'oil, Islam and Israel'. With respect to Islam, the perception exists that the United States harbours a suspicious if not hostile view of Islamic regimes or movements *per se*, and fears the acquisition of nuclear weapons by an Islamic state with more horror than perhaps it would view the possession of a nuclear bomb by others. US tolerance of Israeli nuclear capability, coupled with a policy of combating the proliferation of weapons of mass destruction elsewhere in the Middle East, is taken as reinforcement of this assessment. On the subject of Israel, US analysts are wont to point out that defenders of Israel exercise significant domestic political influence and that this is a fact of life to which critics of US policy in the Middle East must accommodate if they want to be heard. While from the perspective of the United States this may serve as an explanation of its sensitivity on matters relating to Israel, from a Gulf perspective it is an acknowledgment of bias in US policy.

Leaving aside Israel, according to one commentator[5] US policy towards China offers an instructive contrast to its policy towards the Middle East and particularly the Gulf. In its dealings with China the United States does not let the matter of human rights stand in the way of developing a dialogue; its approach is about finding and building on mutual interests. Towards the Middle East, by contrast, the United States does not appear to be open to a process of give-and-take or dialogue; instead it is thought to wield the big stick and act in a divisive manner, playing one country against another, to impose its will. In the face of this, Gulf nationals would like to know how best they might protect their own interests. One possibility raised would be to turn to the Europeans as a potential counterweight to American influence. Certainly, there is sufficient commercial competition between the Western allies to exploit and thereby create some room for manoeuvre. Another way of countering US predominance, perhaps more in keeping with Arab aspirations, would be for the states in the region to develop ties with one another, enhance economic cooperation and build alliances.

On the issue of the contrast between the USA's dealings with China and its posture in the Middle East, from an American perspective it all boils down to a difference in scale. If one calculates the combined GDP of all eight countries around the Gulf – the six GCC states, Iraq and Iran – for a total population of some 110 million, one comes up with a figure approximately equal to the GDP of Switzerland, which has a population of seven million. In other words, the Gulf region lacks the weight needed to impinge on US calculations in the sense that China does. However, this line of reasoning illustrates exactly the nature of the problem the Gulf states face in defining their own agenda. The way the United States

[5] Dr Atif Ajwa.

approaches the area suggests a growing propensity to globalize or internationalize the issues affecting the region.[6] To those in the region it seems that Washington does not permit the Gulf littoral states to develop their own regional agenda from within, but rather that it superimposes its own agenda, without taking into account the subtleties of indigenous politics and cultural differences.

This contention is illustrated, from a Gulf perspective, by the terms of the debate on sanctions on Iraq. Between Americans and Europeans the issues are discussed only within a narrow legalistic framework, focusing on the wording of certain articles in UN Security Council Resolution 687. There is no room in the paradigm for the much wider questions about the meaning and implications of the case of Iraq. Meanwhile, the United States and the Europeans are at least consulting each other. Neither of them is consulting the regional states on what they need, want and aspire to or how they see the situation. For the latter there is no sense of reciprocity and a perception exists that the agenda is being defined for them without regard to regional dynamics.

In response to this set of charges, Gary Sick would conced that the United States probably would not welcome the emergence of a new regional agenda not of its own choosing. However, from mid-1997, certain developments have obliged Washington to re-examine its policies. The rapprochement between Saudi Arabia and Iran has certainly influenced US thinking; the lead taken by the UAE in calling for the rehabilitation of Iraq has also been noted. In the light of Arab reactions to the stand taken by the United States on Iraq in February 1998, it is to be expected that Washington will not relish finding itself subject to the same level of criticism in the region in future.

The lack of reciprocity between the United States and the Gulf states is nonetheless likely to persist in so far as there is little provision for real consultation in the policy-making process in Washington. As noted by one commentator in the Emirates, the executive arm of the US government has been somewhat overtaken by voices in Congress determined to put their stamp on foreign policy, and it is these elements that are wedded to the policy of containment on Iraq and Iran.

Returning to the broader issues of oil politics and the Middle East, a

[6] Dr Jassim Abdulghani Al Khulufi.

note of caution was sounded by one commentator at the seminar in Abu Dhabi, who pleaded for history not to be forgotten and the blame for current woes not to be placed solely at the door of the United States. This analyst[7] had witnessed the 1973 Arab–Israeli war and ensuing oil crisis from both Egypt and France. The winter which followed the war was a particularly cold one in France and the implications of the oil price hike were therefore felt that much more profoundly. How the French felt was made clear during the course of a French television interview of the oil ministers of Saudi Arabia and Algeria, Ahmed Zaki Yamani and Abdel Salam Beleid, respectively. After an hour of discussion, one of the interviewers, the representative of *Le Figaro*, literally shouted at Mr Yamani: 'But Mr Minister, in Europe we are the victims!'[8] According to the raconteur of this incident, neither Europeans nor Arabs have fully absorbed the lessons of 1973, since both continue to wait on each other's actions.

Until this situation changes, it was suggested they had better shoulder some blame for their own circumstances, rather than load it all onto the United States. Meanwhile, perhaps it would be wise to remember that the United States has encountered considerable risks in pursuit of regional security.

Iran and Iraq

As already indicated, the view from the UAE is that discussion in the West about Iraq pays far too little attention to the social and cultural forces at work there. As recounted by one commentator,[9] there has been a fixation with weapons of mass destruction, and neglect of the more dangerous phenomenon of the steady erosion and decimation of the Iraqi middle class. A whole value system is being destroyed and the society turned into a breeding ground for disillusion, terrorism and extremism: 'A politics of despair is taking over the whole nation.' This unquantifiable, invisible dimension of the situation is barely talked about in the security debate, but it actually represents the real long-term threat to

[7] Mustafa Soheim.
[8] Personal recollection of Mustafa Soheim.
[9] Dr Jassim Abdulghani Al Khulufi.

national security in the region. The US agenda, as seen from the Gulf, does not reflect social and cultural considerations, and in any case US policy reflects America's domestic problems. It is the Gulf states, however, that will bear the consequences most directly.

Given what is happening on the ground in Iraq, the view from the UAE is that it would be nonsense to imagine that whoever replaces Saddam Hussein could be some sort of Jeffersonian democrat. Even supposing that the United States could find a democratic alternative to the current Iraqi government, Gulf commentators do not believe that the Iraqis will accept a leadership chosen and nurtured by the United States. Meanwhile, it seems unlikely that someone as close to the regime – and hence implicated in its atrocities – as, for example, Hussein Kamel would be acceptable to anyone. Yet any attempt to wipe out the inner circles that support Saddam Hussein in order to make a clean start would undermine the very security apparatus which has held the government in place. Given the fragmentation of Iraqi society over recent years, with divisions deepening between tribal, ethnic and sectarian groupings, an unravelling at the centre could presage civil strife. Meanwhile, if a new leadership does emerge from within, it will be the product of a society in trauma, united perhaps only around the cause of anti-Americanism and revenge. According to one Gulf commentator, the analogy of what happened in Germany as a legacy of the Versailles treaty may be applicable. By focusing on weapons of mass destruction in Iraq, the international community may be guilty of 'failing to see the forest for the trees'.[10]

In discussion about the confrontation between Iraq and the United Nations weapons inspectors which occurred in late 1997, one commentator[11] noted that the issue for Iraq was the role of Americans on UNSCOM. If the main objective was to have the inspections proceed, it was suggested, and if removal of American participation could ensure this, then surely that should be considered favourably. An American response to this was cautious to dismissive, on the grounds that Iraq could not be permitted to dictate the composition of UNSCOM. However, it may be worth noting here in retrospect that when confrontation loomed again in August 1998, UN Secretary General Kofi Annan indicated that

[10] Idem.

[11] Frauke Heard-Bey.

the way forward would be to take account of Iraqi concerns as well as the principles guiding the inspections.

With respect to Iran, a number of areas of concern emerged in the seminar in Abu Dhabi. The possibility that Iran might find common cause with Iraq was considered. In the view of Edmund Herzig, there were signs that Iran was prepared to establish lines to Baghdad, but only for strictly limited purposes. Only if the United States were to increase pressure on Iran did he think that this pragmatic approach could grow into something more significant. For his part, Raad Alkadiri thought it would be impossible to override traditional suspicions completely, even if the idea of a strategic relationship between Syria, Iraq and Iran might seem to have some appeal for Baghdad as a way out of its current isolation.

Another subject of interest was the rapprochement between Iran and Saudi Arabia and the implications of this for the UAE.[12] According to Edmund Herzig, Iran is likely to want to develop better relations with all the GCC states, not with one or another at the expense of the rest. He bases this assessment on the lessons of Iran's policies in Central Asia and the Caucasus, where Tehran has sought to foster an even-handed approach rather than a divisive one. Meanwhile, the experience of Armenia may be instructive. Even though US policy was designed to isolate Iran, Armenia managed to keep its access in Washington at the same time as pursuing friendly relations with Iran, and it continued to be the recipient of US aid. With respect to Iran, Dr Herzig suggested that the Arab Gulf states might best proceed by identifying ways in which they could engage Iran to mutual benefit and concentrate on those, rather than on the areas of contention.

The new generation and policy priorities

The discussions in Abu Dhabi in November 1997 revealed much interest in Saudi Arabia's problems over finding employment for the new generation in the Kingdom. Even though, at the time of the seminar, the extent of the oil price slump that was to erode government revenues in the coming months was not apparent, the relationship between revenues, welfare provision and job creation was a matter of concern.

[12] Dr Hasan Alkim.

It is clear that Saudi Arabia shares certain features with the other member states of the GCC, but the parallels should not be overdrawn. Each of these states has unique characteristics. Within the context of the UAE, however, there is an impetus to address simultaneously, as policy issues, the questions of expectations for oil revenues; employment prospects; the place and size of the expatriate labour force; and the quality and content of education.

In short, the citizens of the UAE wish to concentrate their attention on a number of issues. Their concerns could be considered commensurate with the concerns of citizens in other countries around the world caught up in the imperatives of 'globalization'. However, as demonstrated above, there exists a sense that powers outside the region have defined the agenda of globalization and depicted the Gulf region within this context, without paying much attention to the political and cultural features of the societies themselves. A new shake-up in the regional power balance looks set to demonstrate that any order imposed from outside, without sufficient heed to the dynamic within, cannot be sustained for long.

As of this writing, there is reason to believe that the period of US predominance in the management of Gulf security may have peaked. That is not to say that the United States is likely to withdraw its forces altogether from the region, but it may be looking for a less high-profile approach and prefer to rely more on an 'over the horizon' task force and missile capability. Meanwhile, as demonstrated in the discussion which took place in Abu Dhabi in November 1997, the desire of Gulf Arabs to build up regional alignments and connections in response to specifically regional concerns is clearly present. These sentiments seem likely to portend an end to the relative isolation of Iraq and Iran. Even so, the reintegration of both will only happen gradually and most probably not without some new crisis points along the way.